D1187659

RFID and Contactless
Smart Card Applications

RFID and Contactless Smart Card Applications

Dominique Paret

Translated by
Roderick Riesco, MA
Member of the Institute of Translation and Interpreting, UK

John Wiley & Sons, Ltd

Other Wiley Editorial Offices

John Wiley & Sons Inc., 111 River Street, Hoboken, NJ 07030, USA

Jossey-Bass, 989 Market Street, San Francisco, CA 94103-1741, USA

Wiley-VCH Verlag GmbH, Boschstr. 12, D-69469 Weinheim, Germany

John Wiley & Sons Australia Ltd, 42 McDougall Street, Milton, Queensland 4064, Australia

John Wiley & Sons (Asia) Pte Ltd, 2 Clementi Loop #02-01, Jin Xing Distripark, Singapore 129809

John Wiley & Sons Canada Ltd, 22 Worcester Road, Etobicoke, Ontario, Canada M9W 1L1

Wiley also publishes its books in a variety of electronic formats. Some content that appears in print may not be available in electronic books.

Library of Congress Cataloging-in-Publication Data:

Paret, Dominique.
 [Applications en identification radiofréquence et cartes à puces sans contact. English]
 RFID and contactless smart card applications / Dominique Paret ; translated by Roderick Riesco.
     p. cm.
 Translation of: Applications en identification radiofréquence et cartes à puces sans contact.
 Includes bibliographical references and index.
 ISBN-13 978-0-470-01195-9 (HB)
 ISBN-10 0-470-01195-5 (HB)
 1. Smart cards. 2. Radio frequency identification systems. I. Title.
 TK7895.S62P3713 2005
 621.384′19–dc22

                                                    2005005159

British Library Cataloguing in Publication Data

A catalogue record for this book is available from the British Library

ISBN-13 978-0-470-01195-9 (HB)
ISBN-10 0-470-01195-5 (HB)

Typeset in 10/12pt Times by Laserwords Private Limited, Chennai, India
Printed and bound in Great Britain by Antony Rowe Ltd, Chippenham, Wiltshire
This book is printed on acid-free paper responsibly manufactured from sustainable forestry in which at least two trees are planted for each one used for paper production.

Contents

Acknowledgements

This new field of "contactless" Radio Frequency IDentification (RFID) is very active, and many skilled people are working in it. I have been lucky enough to meet many of them on numerous occasions, and therefore it is very difficult for me to acknowledge all of them individually.

At the risk of some unfairness, therefore, I should like to offer my special thanks to colleagues at Philips Semiconductors who run the multidisciplinary Contactless Identification teams at Hamburg, Germany and at Gratkorn, near Graz in Austria, with whom I have had the pleasure of working in this area for many years. In the first place, I would like to mention

Reinhard Meindl
Steffen Drews
Wolfgang Tobergte
Jurgen Nowottnick
Holger Kunkat
Thomas Giesler
Stefan Posch
Michael Jerne
Wilfried Pestchenig
Thomas Rudolph
Franz Amtmann
Huber Watzinger
Christian Schwar

and, finally, Martin Bührlen, for the numerous documents and photographs that he has been kind enough to provide me with.

Lastly, I should be ungrateful if I failed to thank the numerous professional colleagues, "friends and rivals", whom I encounter regularly at AFNOR and ISO meetings (they will know who they are), for their remarks and comments on the production of this book; it is thanks to them that this field of RFID will see its eagerly anticipated development.

I should like to end by offering my very sincere thanks to Mrs Sylvie Bourgeois, who was responsible for proof-reading, checking, correcting and standardizing the parameters and equations in this book.

Dominique Paret
Meudon

Foreword

The first RFID systems using a silicon microchip were developed well over ten years ago. Almost unnoticed by the public, early applications in animal identification or car immobilization have been used successfully for more then a decade now. Even the first large contactless ticketing applications, started around 1996 in Asia, have not earned a lot of attention outside the technical press.

Now, in the mid-2000s, the picture has completely changed. The announcement of new and emerging RFID applications, such as the so-called "electronic barcode" to be used on food and non-food products, for example, or the contactless passport and many other applications, never even thought of ten years ago, are widely discussed in the press. Today, the RFID market belongs to one of the fastest growing sectors in the radio technology and IT areas. Also, the increasing number of companies actively involved in the development and sale of RFID systems indicates that this is a technology and a market that should be taken seriously.

Furthermore, in recent years, contactless identification has developed into an independent interdisciplinary field that no longer fits into any of the conventional fields. It brings together elements from extremely varied subjects, such as RF technology, semiconductor technology, telecommunications, manufacturing technology, data protection and cryptography and many other related areas.

In the early years, the RFID implementer and system integrator often suffered from a lack of RFID standards. Almost all products have so far been proprietary to the different companies. Interoperability between products from different sources is not the state of the art. In the meantime, numerous standards have been released and some more are still under development. Current standards cover all of the most important application areas of RFID, such as ticketing, banking, smart labels, electronic passports, animal ID, freight containers and, last but not least, the electronic product code (EPC). Nevertheless, it has not become easier at all. Today's challenge is, in fact, to keep an overview of all of the new emerging standards and new technologies and to be able to select those that best fit one's own needs. Furthermore, there still exists a lack of literature on the technical basis and the standardization framework of RFID. In this spirit, I really appreciate the present work of Dominique, who is a well-known and respectable colleague in RFID standardization work, which I am sure will become an important source of technical basics for contactless smart cards and smart labels.

Klaus Finkenzeller

Preface

"Contactless" radio frequency identification (RFID) is currently a flourishing subject area. Having worked in this field for many years, I felt the need to draw up a "progress report" on this subject. Hitherto, very little basic information or technical training in terms of applications and technology has been available to engineers, technicians and students. We trust that this book will at least partially remedy this deficiency.

The aim of this book is to provide, at a given date, the most comprehensive guide to the technologies and the various methods of applying them. This book is not intended to be encyclopedic, but it is a solid and thorough technical introduction to this subject. It is thorough in the sense that all the significant aspects of "contactless" applications (principles, theories, technology, components, conformity with standards, security, detailed examples of applications, etc.) are dealt with in detail.

Additionally, as an aid to learning, instead of placing all the theories and equations in a specific chapter and risking an information overload, I decided to distribute them through the text, as far as practicable, so that the reader would be able to see the overall picture at every point, covering theory, applications, implementation, technological and economic aspects, and so on.

This book is designed to introduce the reader to real implementations of "contactless" applications and it provides full details of all the complexities of system development.

As this field is developing so rapidly, I expect that I will have to take up my pen again (or rather return to the keyboard) in two or three years' time to bring you the latest news, but I hope that this book will meet the needs of most users for the time being. In any case, it will ensure that solid foundations are laid while we await future developments.

I hope you enjoy this book – may your applications run smoothly!

IMPORTANT NOTE

Readers' attention is drawn to the important fact that, in order to provide full coverage of the field of "contactless" technology, this book describes very many patented technical principles that are subject to licensing and associated rights (bit coding, modulation techniques, collision management devices, technical assemblies, etc.) and which have already been published in official professional texts and communications or at public conferences and seminars but whose use is strictly subject to current regulations.

Introduction

In my previous book, *Identification radiofréquence RFID et cartes à puce sans contact – Principe et description [Radio frequency identification (RFID) and contactless smart cards: Principle and description]*, published by Dunod described the main physical and electrical principles of the operation of "contactless" systems, providing details of the bit coding used, the communication protocols generally employed, the deterministic and probabilistic principles of collision management, the current standards relating to radio interference and security and health aspects, and so on. Consequently, these points will not be covered in detail in this book.

The present book is intended for those readers who already have some grounding in these areas and who now wish to enhance their applied knowledge with a view of carrying out practical work on systems design, and also for all potential and existing users, to enable them to specify in detail the actual performance of the systems they wish to use.

Consequently, in some chapters it will be necessary to provide new supplementary theoretical information, which is "application-oriented" rather than purely theoretical, and to provide a way of approaching the problem of implementing these applications, something which often seems baffling to many potential users.

To conclude this brief introduction, I will outline the level of technical knowledge that the readers of this book are assumed to have.

The book is intended for electronic engineers and technicians, as well as lecturers and students in this field and all enthusiasts interested in new developments, but, despite its apparently highly technical nature, it is also intended for anyone with even the slenderest technical knowledge in the fields of automation, packaging, industrial process monitoring, product traceability and logistics, and so on, since these areas are already feeling the impact of "contactless" technology and RFID, which is beginning to be applied here, either in addition to bar codes or as a replacement for them.

Regardless of their background, readers will find that this book contains an abundance of technical information that, unlike the journalistic jottings offered by non-specialist or semi-specialist periodicals, will give them a solid and professional grounding in the day-to-day problems encountered in the field.

Part One
Review and supplementary information

In an earlier book[1], I introduced the basic elements of "contactless" applications, setting out the fundamental principles of the operation of contactless devices, remote power supply, communication, collision management, security aspects of transmitted messages and matters relating to the current and other standards.

The three chapters forming the first part of this book contain a great deal of important theoretical technical information to supplement the content of the first book, which described the general principles of contactless technology. I preferred to include this information in this second book, since, in my view, although its content is related to general principles, it really comes into its own when we make a much closer examination of the definitions and solutions encountered in applications.

Although the study of any contactless application requires us to change our viewpoint constantly from the transponder to the base station and back again, I have decided to set out the technical details, the review of earlier information and particularly the supplementary information in two separate chapters, the first of which deals with transponders only, while the other is concerned with the problems of base stations.

---

[1] Previously published: *Identification radiofréquence et cartes à puce sans contact – Description*, by Dominique Paret, Dunod, 2001 – ISBN 2 10 005955 6.

1
Review

Before going into the details, let us briefly review how the components of an RFID system are organized and how they operate. *Figures 1.1a* and *1.1b* show the structure of a system of this kind.

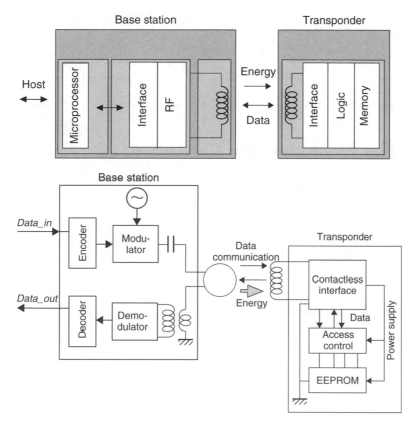

Figure 1.1a Block diagram of a contactless system

Note: for more detailed information, you should refer to the author's very comprehensive book on the subject, published previously by Dunod.

RFID and Contactless Smart Card Applications   D. Paret
© 2005 John Wiley & Sons, Ltd

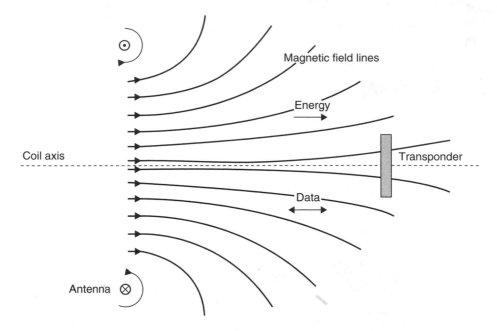

Figure 1.1b   Operating principle of a "base station–transponder" system

1.1   The Elements of a Contactless Device

1.1.1   The transponder

Let us start with the transponder, which uses a remote memory (ROM, PROM, E2PROM, etc.) to store data for the application in question, and which also controls communication and contains the part used for the RF link.

1.1.2   The air interface

The air forms the communication medium. The electromagnetic RF waves carry the data. The air also plays a part in the (magnetic) coupling between the antennae of the transponder and the base station.

1.1.3   The base station

The base station comprises an analogue part used for receiving and transmitting RF signals, the circuits controlling the protocol for communication with the transponder, the communication management (management of possible collisions, authentication, cryptography, etc.) and finally an interface for dialogue with the host system.

1.1.4 The host system

Finally, to complete this brief survey, we have the host system, providing the higher level control of the application.

Now that the general picture is in place, let us examine each of these elements in more detail.

1.2 General Operating Principles of the "Base Station–Transponder" Pair

For applications in which the remote element – the transponder – has no on-board power supply and therefore operates with a remote power supply, *Figures 1.2a* to *1.2c* provide a summary of the physical principles of the three main stages of operation of a "base station–transponder" system.

1.2.1 Transfer, energy supply and remote power supply

An alternating electrical signal at radio frequency, called the "carrier", creates a magnetic field, by means of the winding forming the base station antenna, to remotely induce a voltage in the antenna of the transponder; this voltage is rectified and filtered to provide a local power supply for the transponder (see *Figure 1.2a*).

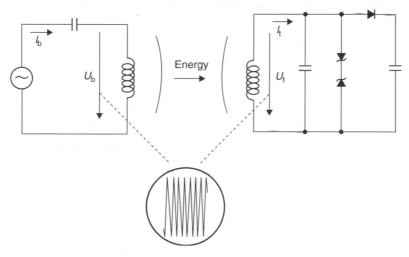

Figure 1.2a   Energy transfer phase

1.2.2 Data transfer from the base station to the transponder: the "uplink"

In order to transmit commands and data to the transponder, the carrier is modulated with a binary stream based on specific bit codings (see *Figure 1.2b*).

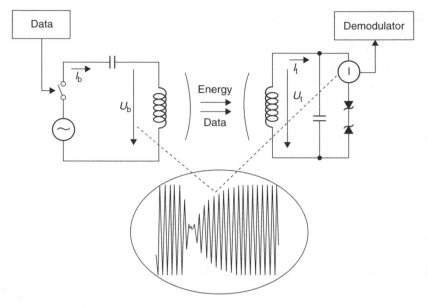

Figure 1.2b   Uplink phase

1.2.3   Data transfer from the transponder to the base station: the "downlink"

The transponder can communicate with the base station by means of a modulation carried out by varying the load that it represents in the field (see *Figure 1.2c*).

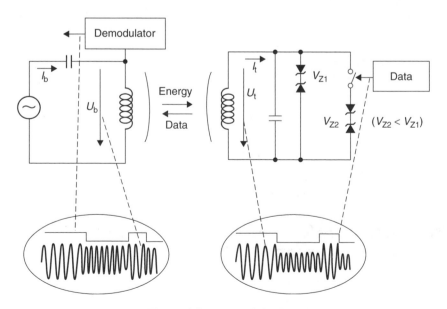

Figure 1.2c   Downlink phase

To explain the operation of the "base station–transponder" pair more simply, I shall look inside most of its constituent parts and try to explain the general principles as clearly as possible, breaking the process down into two main sections, namely, the transfer of energy on the one hand and the operation of the uplink and downlink on the other.

1.2.4 Energy transfer

The transponder can only operate if it is supplied with power. In the present case, this is done by the remote power supply principle, which will be described in detail in this section.

Figure 1.3 shows how the energy transfer is carried out by means of an alternating magnetic field $H(t)$ produced by the flow of an alternating current $I(t)$ in the antenna of the base station, according to the general equation

$$H = NI,$$

where N represents the number of turns per metre of the antenna carrying the current I. The strength of the magnetic field H is expressed in amperes per metre (A/m).

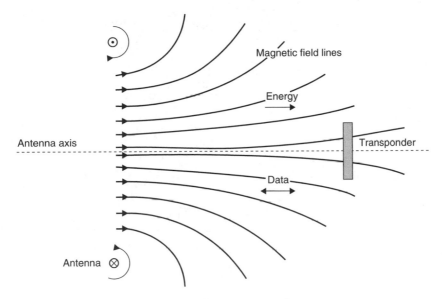

Figure 1.3 Energy transfer

In the medium in which this magnetic field develops, there is a corresponding associated induction $B(t)$ having the general form

$$B(t) = \mu \, H(t),$$

where μ represents the magnetic permeability of the medium(s) through which the field passes. This magnetic permeability is expressed by

$$\mu = \mu_0 \mu_r,$$

where μ_0 represents the permeability of the air ($\mu_0 = 4\pi \times 10^{-7}$ henrys/metre) and μ_r is the relative permeability of the medium concerned (water, metal, etc.). The value of the magnetic induction B is expressed in teslas (T).

The Biot–Laplace and Biot–Savart laws establish the relationship between the current I flowing in an element dl of an electrical circuit having a length l and the intensity of the magnetic induction B as a function of the distance x between the measurement point (of unit vector $\vec{u}$) and the element dl producing the field according to the formula:

$$\overrightarrow{dH} = \frac{(I \cdot \overrightarrow{dl}) \wedge \vec{u}}{4\pi (x^2)}$$

Therefore,

$$\vec{B} = \mu \vec{H} = \frac{(\mu I)}{4\pi} \oint \frac{\overrightarrow{dl} \wedge \vec{u}}{x^2}$$

where $\oint$ is a line integral calculated over the whole length l and $\wedge$ is the notation of the vector product.

$$\overrightarrow{dH} = \frac{(I \cdot \overrightarrow{dl}) \wedge \vec{u}}{4\pi x^2} \Rightarrow dH = \frac{I \cdot dl \cdot \sin \alpha}{4\pi x^2}$$

Now,

$$\alpha = \frac{\pi}{2} \Rightarrow \sin \alpha = 1$$

and therefore

$$dH = \frac{I \cdot dl}{4\pi x^2}$$

In the axis of the coil, we find

$$dH_x = \frac{I}{4\pi x^2} dl \sin \theta$$

Now,

$$x = \frac{r}{\sin \theta}$$

and therefore

$$dH_x = \frac{1}{4\pi r^2} \sin^3 \theta \, dl$$

On the other hand, at a distance d in the axis of the coil, (see *Figure 1.4*)

$$x = \sqrt{r^2 + d^2}$$

and

$$\sin^3 \theta = \frac{r^3}{x^3}$$

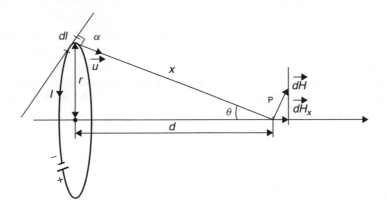

Figure 1.4 Magnetic field strength as a function of distance and antenna radius

and therefore

$$\sin^3 \theta = \frac{r^3}{(r^2 + d^2)^{3/2}}$$

This gives us

$$dH_x = \frac{1}{4\pi} \cdot \frac{r}{(r^2 + d^2)^{3/2}} dl \Rightarrow H_x = \frac{I}{4\pi} \cdot \frac{r}{(r^2 + d^2)^{3/2}} \int_0^{2\pi r} dl = \frac{Ir^2}{2(r^2 + d^2)^{3/2}}$$

Solving this equation for a circular base station antenna with the radius r having a flat configuration, with N turns, at a point P located at a distance d along the axis of the antenna, we find:

$$B(d, r) = \mu \frac{r^2}{2[(r^2 + d^2)^{3/2}]} NI = \mu \cdot H(d, r)$$

Examples, in air, with $\mu = \mu_0 = 4 \times 3.14 \times 10^{-7}$:

— if $H = 1$ A/m, then $B = \mu_0 H = 1.256\,\mu\text{T}$;

— if $B = 1\,\mu\text{T}$, then $H = \dfrac{B}{\mu_0} = 0.796$ A/m.

NOTE

To ensure that the following chapters are understood, it will be useful to review the following concepts, which are in fact simply different aspects of the equation shown above.

$B(d)$ at $r = $ constant

It is useful to examine the variations of magnetic induction as a function of the distance between the base station antenna and the transponder, since there are numerous applications for which the radius of the base station antenna is constant and the distance between the base station and the transponder is variable (e.g. ticket presentation, access control, object identification, etc.). We shall therefore examine in detail the general equation of magnetic induction,

$$B(d, r) = \mu \frac{r^2}{2[(r^2 + d^2)^{3/2}]} NI$$

assuming in this case that the value of r is constant. It is therefore a function $B(d)$.

To investigate its variations, we calculate its derived function $B'(d)$, which is of the type $-v'/v^2$. After expansion and reduction, we have

$$B'(d) = \frac{-3(\mu NI)r^2}{2} \cdot \frac{d}{(r^2 + d^2)^{5/2}}$$

This first derivative $B'(d)$ is cancelled for $d = 0$, and $B(d, r)$ has a maximum with the value

$$B(d = 0) = B_{\text{max}} = \frac{(\mu NI)}{2r}$$

Now we calculate the second derivative $B''(d)$ to determine whether any points of inflection are present. This derivative is of the type $(u'v - v'u)/v^2$. After expansion and reduction, we have

$$B''(d) = \frac{-3(\mu NI)r^2}{2} \cdot \frac{r^2 - 4d^2}{(r^2 + d^2)^{7/2}}$$

This second derivative $B''(d)$ is cancelled for $4d^2 = r^2$, signifying that the curve of the variations of $B(d)$ therefore has two symmetrical points of inflection for $d = \pm(r/2)$.

$$B\left(d = \frac{r}{2}\right) = \mu \frac{1}{2r\left[\left(\dfrac{5}{4}\right)^{3/2}\right]} NI = B(d = 0) \frac{1}{\left(\dfrac{5}{4}\right)^{3/2}} = \frac{B(d = 0)}{1.397}$$

The curve of the variations of $B(d)$ for a constant radius of the base station antenna is shown in *Figure 1.5*.

In this figure, these variations are deliberately shown with respect to two different radii of the antenna, r_1 and r_2, where $r_2 > r_1$, to illustrate the content of the following text.

Still considering $B(d)$ for a constant radius, we shall now examine two common cases of application.

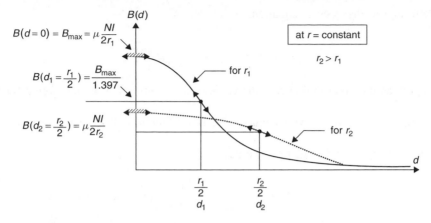

Figure 1.5   Variation of $B(d)$ with constant radius of base station antenna

Operation of the transponder at a specified distance

If the transponders have to operate at a specified distance d_1, it will be necessary to produce a closely specified magnetic induction $B(d_1)$, which, having a constant product (NI), will be obtained for a specified value of r_1.

Operation of the transponder in a specified range of distances

We shall now examine the case in which transponders are required to operate, not at a fixed distance from the base station but within a specified range of distances. In this case, to obtain the greatest possible reproducibility of operation, it is generally desirable for the transponder to be exposed as far as possible to a magnetic induction B that is, "virtually" constant over the desired range of distances. To do this, as indicated in the figure, the curve representing the variations of magnetic induction B as a function of the distance d must be as flat as possible over the range of distances from d_1 to d_2.

In this case, still with a constant product (NI), it is preferable to increase the value r of the radius of the antenna, since, even if the maximum induction decreases as a result, in the range of distances d_1, d_2, its variations change more slowly as a function of the distance. In this type of application, the designer must therefore calculate the literal expressions of $B(d_1)$ and $B(d_2)$ – with r as the parameter – to enable him to determine the exact value of r that minimizes the difference $\Delta = B(d_1) - B(d_2)$.

We should note in passing that the "delta" (Δ) equation takes into account the variables r, d_1 and d_2 (where $d_2 = d_1 + \text{const}$) and that at a desired value of Δ the value of r will depend on d_1 and on const, in other words on the two parameters "initial minimum distance and desired range of distances", and not solely on the value of the desired range.

$B(r)$ at $d = \text{constant}$

It is also useful to examine the variations of $B(r)$ at $d = \text{constant}$, since this representative case is found in many applications where the operating distance between the base station and transponders is constant (e.g., in the identification of objects moving on a conveyor in a production line).

We shall now examine the general equation for magnetic induction, to find the optimal value for the radius of the base station antenna for this kind of application.

Returning to the general equation,

$$B(d,r) = \mu \frac{r^2}{2[(r^2 + d^2)^{3/2}]} NI$$

let us now assume that in this case the value of d is constant. It is therefore a function $B(r)$.

To study its variations, we calculate its derived function $B'(r)$, which is of the type $(uv' - v'u)/v^2$.

After expansion and reduction, we have

$$B'(r) = \frac{\mu NI}{2} \cdot \frac{r(2d^2 - r^2)}{(r^2 + d^2)^{5/2}}$$

This derivative is cancelled for $r = 0$ and for $r = \pm d\sqrt{2} = \pm 1.414d$.

Examining the changes of signs in this derivative, we see that $B(r)$ passes through a minimum at $r = 0$ (which was physically predictable, since $B(r = 0) = 0$!) and through a maximum for $r = \pm 1.414d$. The value of this maximum is

$$(r = 1.414d) = \mu NI \frac{1}{d\sqrt{27}}$$

The curve of the variations of $B(r)$ at the distance $d = $ constant is shown in *Figure 1.6*.

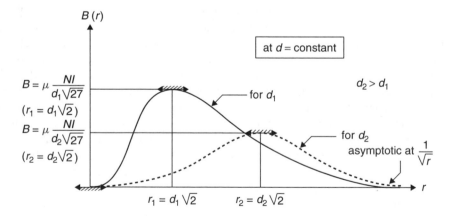

Figure 1.6   Variation of $B(r)$ with constant distance d

Now you know what needs to be done to optimize a system operating at a fixed distance. Everything else being equal, if you specify $r = 1.414d$ you will obtain the greatest induction; alternatively, for a given induction, you simply decrease the value of the current I to obtain it.

Rationalized equation of induction $B(d, r)$ To generalize the argument concerning the relative values of d and r, we shall now rationalize this general magnetic field strength or magnetic induction equation. To do this, we assume that $d = a \cdot r$. The general equation then takes the form

$$B(a,r) = \mu \frac{1}{[(1+a^2)^{3/2}]} \cdot \frac{NI}{2r} = \mu \cdot H(a,r)$$

Now, the value $NI/2r$ represents the value of $H(a,r)$ for $a = 0$, that is, $H(0,r)$ in the centre of the flat circular antenna, and therefore

$$H(a,r) = \frac{1}{(1+a^2)^{3/2}} H(0,r)$$

We can now look at some specific examples according to the relative values of r and d or values of the ratio $a = d/r$.

Table 1.1 shows, for a given value of r, the values of the correction factor $(1+a^2)^{3/2}$ and its reciprocal as a function of the ratio $a = d/r$.

Table 1.1

d	a	$H(0,r)/H(a,r) = (1+a^2)^{3/2}$	$H(a,r)/H(0,r) = 1/(1+a^2)^{3/2}$
0	0	$\sqrt{1} = 1$	$= 1$
$0.5r$	0.5	$\sqrt{1.953} = 1.397$	$= 0.716$
r	1	$\sqrt{8} = 2.828$	$= 0.354$
$1.33r$	1.33	$\sqrt{21.34} = 4.62$	$= 0.216$
$1.414r$	1.414	$\sqrt{27} = 5.196$	$= 0.192$
$1.5r$	1.5	$\sqrt{34.33} = 5.86$	$= 0.170$
$1.66r$	1.66	$\sqrt{53.8} = 7.34$	$= 0.136$
$2r$	2	$\sqrt{125} = 11.8$	$= 0.0847$
$3r$	3	$\sqrt{1000} = 31.62$	$= 0.0316$
$4r$	4	$\sqrt{4913} = 70$	$= 0.0143$
$4.5r$	4.5	$\sqrt{9595} = 98$	$= 0.0101$
$5r$	5	$\sqrt{17576} = 132$	$= 0.0076$

Very near field ($d \ll r$, or $a \ll 1$)   If the distance d between the base station antenna and the transponder is very small compared with the radius r ($d \ll r$, that is, has a very small value of a), the transponder is in the "very near field" of the transmitting antenna and the equation takes a form of the type const/r and the induction $B(d)$ changes to $1/r$.

Note that for $d = 0$ (and $a = 0$), we return to the well-known formula for induction in the centre of a flat circular coil:

$$B(0,r) = \mu \frac{NI}{2r} = \mu \cdot H(0,r)$$

Near field (d of the order of r, or a of the order of 1)   The special case where $d = r$ ($a = 1$) is frequently encountered in contactless applications. In this case, the value of $B(d,r)$ becomes

$$B(d,r) = \mu \frac{1}{2r\sqrt{8}} NI = \frac{B(0)}{\sqrt{8}}$$

Near field (d > r, or a > 1)   If the distance d is large (but not enormously large!) with respect to the radius of the antenna a (e.g. $a = 2, 3, \ldots, 5, \ldots$), the transponder is still in the "near field" of the transmitting antenna. The value of a^2 is therefore large with respect to 1, and the equation is reduced to the form

$$B(d, r) = \mu \frac{1}{2r \left(\dfrac{d^2}{r^2} \right)^{3/2}} NI = \mu \frac{1}{2r \dfrac{d^3}{r^3}} NI$$

and therefore

$$B(d, r) = \mu \frac{r^2}{2 \times d^3} NI$$

and for a specific constant value of r it takes a form of the type const'/d^3. The induction $B(d)$ then changes to $1/d^3$.

By way of a summary, *Figure 1.7* shows the trend of the variations of the general function $B = f(d)$, in the very near and near fields, where r is constant.

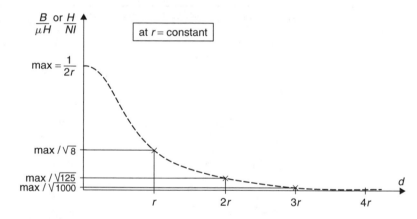

Figure 1.7   General variation of $B(d)$ (near and far fields)

Far field (d ≫ r, or a ≫ 1)   In Chapter 6 we shall examine in detail what are known as far (or very far) fields, in other words those in which $d \gg r$, which, as we will show, are no longer subject to the Biot and Savart law but relate to the standard propagation of radio waves.

In this case, the predominance of the field H is reduced, and radiated fields E and H are present.

1.2.5   Remote power supply

The principle of the remote power supply to the transponder will now be summarized.

The magnetic induction B causes the appearance of a magnetic flux Φ. At a distance d, in a conductor having a total cross section $S = N \cdot s$ (where N is the number of turns), this flux is equal to $\Phi = B(d) \cdot S$.

The instantaneous variation of magnetic flux causes the appearance of an induced potential difference (pd), $u(t)$, across the terminals of the conducting element acting as the receiving antenna, which is affected by the variation of the flux according to the well-known equation

$$u(t) = -\frac{d\Phi}{dt}.$$

That is the essential description of the appearance of a pd across the terminals of the transponder antenna, corresponding to an induction B present in the vicinity of the transponder, generally ranging from a few tens of nanoteslas (nT) to a few hundred microteslas (μT).

When the transponder has been tuned to the carrier frequency by means of tuned resonant circuits of the LC type, a voltage (derived from this induced voltage) of the order of several volts will be applied to the terminals of the integrated circuit.

The next step is to rectify this incident-induced alternating sinusoidal voltage, to recover a unidirectional voltage, after which it must be filtered and regulated to provide a continuous voltage, using filter capacitors whose values are, evidently, functions of the received carrier frequency (e.g., of the order of a nanofarad (nF) for 125 kHz, or several tens of picofarads (pF) for 13.56 MHz).

The quality and quantity of the energy transfer depend on the frequencies to which the two antenna circuits (base station and transponder) are tuned; their precision, tolerances, fluctuations and so on; the coupling coefficient (distances, geometrical shape of the antennae, etc.); and, of course, the quality factors of the tuned circuits of the base station and transponder.

However, if everything else is equal, the critical value – the threshold – for a good supply is achieved and corresponds on the one hand to a certain minimum induction B_{min} but also, on the other hand, to a particular magnetic coupling (which of course depends on the distance, among other factors) related to the possibility of providing a sufficient quantity of energy, which will be denoted k_e.

1.2.6   Communication between the base station and transponder

We shall now look at the exchanges that have to take place between the base station and transponder, and more precisely the exchanges from the base station to the transponder, which will be referred to as the "uplink", and those from the transponder to the base station, known as the "downlink".

NOTE
The technical literature provides many definitions of the "direction" of the "uplink" and "downlink". In this book, it is simply assumed that the base station system supplying energy to the transponder is the reference point of the system, and that the links originating from the base station are "uplinks".

The uplink   In the upward direction, the base station communicates with the transponder by means of a coding (binary) and a carrier modulation system that must not affect (or must only slightly affect) the quality of its remote power supply function.

In this case, the value of the coupling k_e is sufficient to provide the remote energy and power supply to the transponder, and also to provide the uplink for the communication between the base station and the transponder.

The downlink   In the downward direction, from the transponder to the base station, the transponder must be able to communicate with the base station. The technical principle implemented in most commercially available transponders is that of "load modulation". The most widely used principle is that of a type of modulation based on a variation of resistive load.

By varying this load, the transponder modifies the energy consumption that it represents in the magnetic field and, because of the magnetic coupling between the transponder and the base station, tends to modify the current flowing in the circuit of the base station antenna.

Minimum coupling   I have written "tends to modify", rather than "modifies". This is because, in order for this modulation/load variation of the transponder to be sensed and detected in the base station circuits used for this purpose, it is necessary for a minimum magnetic coupling $k_{\text{application}}$ to be present between the transponder and the base station antenna. We must distinguish between the concept of the coupling required for the energy transfer k_e for the remote power supply of the transponder and the coupling required to establish the downlink to enable the application to operate correctly $k_{\text{application}}$.

1.3   Before We Continue . . . Conventional Notation

Before we continue, let me summarize the conventional notation used throughout the book:

- subscript 1 for everything relating to the base station (the "primary" physical event);

- subscript 2 for everything relating to the transponder (the "secondary" physical event);

- a lower-case s for everything relating to "serial" physical representations;

- a lower-case p for everything relating to "parallel" physical representations;

- lower-case letters for instantaneous values or values of the complex variable;

- upper-case letters for values of permanent or rms parameters.

NOTE
In order to avoid cumbersome arrays of symbols when explaining specific points, I have simplified the notation (which is still adequate, and certainly much clearer).

2

The Transponder: Supplementary Information

Very often, the choice of a transponder forms the starting point for the development of an application. This choice is generally based on the performance (architecture and capacity of the memory, collision management device, security aspects, etc.), the technology and other matters.

My previous book included a full description of the kinds of performance that could be expected from transponders. I shall now briefly review the various options in terms of technology.

Clearly, the simplest approach is to find a complete transponder, including its integrated circuit and antenna, as a ready-made commercial product, listed in a manufacturer's catalogue; in this case it is simply necessary to check its specifications which, with a few exceptions, are generally very sketchy. Few manufacturers of transponders (incorporating integrated circuits and antennae) provide clear details of the magnetic induction (or even the magnetic fields in air), the quality factors, the bandwidths, the minimum and maximum coupling required for satisfactory operation, and so on.

If the user decides to produce his own transponder (by selecting or developing his own integrated circuit, special configuration, antenna, etc.), he must first examine the characteristics of the integrated circuits and then move on to the design of the transponder antenna.

2.1 Ready-made Products

Many companies offer ready-made transponders. In theory, the end user merely has to scan a catalogue and choose the most appropriate transponder for his applications.

However, this is not such a simple matter, because these general-purpose catalogues are rather abbreviated and provide very little in the way of practical technical information for use in applications, apart from a few conventional principal parameters such as the operating frequencies, the memory capacity, and so on.

RFID and Contactless Smart Card Applications   D. Paret
© 2005 John Wiley & Sons, Ltd

The user often needs to investigate the minute details of the electrical and magnetic characteristics of these products, for example, by examining and requesting quantitative information on the tuning frequency of the transponder, rather than just the frequency at which it is supposed to operate, its quality and coupling factors, the bandwidth at different levels of magnetic field, the threshold magnetic field value, the precise description of the method of managing possible collisions and the time values associated with this management, as well as the tolerances of all kinds, the temperature coefficients, and so on – simply to be sure that his application will eventually work.

If this is successful, we must first congratulate the manufacturers of transponders sold as ready-made products; following this process, the user will have ready access to most of the parameters required to design a truly reliable application.

2.2   Specification or Choice of the Transponder Integrated Circuit

Regardless of the range, the local magnetic field strength or the magnetic induction present in the immediate environment of the transponder, the integrated circuit incorporated in it will only operate correctly if an alternating sinusoidal voltage V_{ic} (matching the value shown on the integrated circuit data sheet) is applied to its two input terminals to ensure that it performs all of its functions.

Conventional specifications of an integrated circuit for use in a transponder   The data sheet of an integrated circuit for a transponder generally comprises two major sections:

- the first describes the functionality of the component, that is, its structure, its memory capacity, the structure of the memory, its communication protocol, the methods used for collision management, the operation of its protection systems, and so on;

- the second is more physical, describing the electrical part of the integrated circuit, which generally needs only two inputs, A and B, to operate. This section is frequently fairly brief and consists of a few parameters such as

  - minimum, nominal and maximum operating voltage V_{ic}rms;

  - input capacitance C_{ic};

  - input resistance R_{ic};

  - power consumption P_{ic}rms;

and a few additional graphs relating to these values. Not much to go on, but we must do our best.

IMPORTANT NOTES
You should note that, if these values are to have any meaning, it is essential

- that the values of C_{ic} and R_{ic} should be measured with a high-performance network analyser such as a Hewlett Packard/Agilent model (this is not an advertisement!), and defined for one or more specified sinusoidal alternating voltages V_{ic};

- that the variations of these values should be known as a function of the voltage V_{ic} applied to the integrated circuit. This particular point will be examined in detail subsequently;

- that there are clear details of how the power consumption of the integrated circuit is calculated, for example, by means of the formula:

$$P_{\text{ic}} = \frac{V_{\text{ic}}^2}{R_{\text{ic}}}, \quad \text{in watts RMS}$$

that is, as an alternating (AC) and not a continuous (DC) value, as is sometimes the case. The latter situation is often misleading, since integrated circuits are frequently synchronized with the incident frequency originating from the base station, and therefore consume no power in the absence of the clock: this has nothing in common with the real application.

Having introduced these considerations, let us see how these integrated circuits can be used to construct high-performance transponders. To facilitate the understanding of the operation of the system, I shall break down the complete structure of the transponder and explain the following elements in turn:

- the electrical aspects of the transponder antenna only:
  - in its series representation;
  - in its parallel representation;
- the electrical aspects of the transponder antenna only, tuned, outside the magnetic field:
  - equivalent impedance;
- the electrical aspects of the transponder antenna, tuned and loaded, outside the magnetic field:
- the electrical circuit of the system;
- the electrical aspects of the transponder antenna, tuned and loaded in the presence of a magnetic field:
  - power matching;
  - value of the equivalent voltage generator,
  - voltage present at the terminals of the transponder integrated circuit.

2.3   The Transponder Antenna

We shall examine in detail how to define the essential characteristics of the transponder antenna and its numerous equivalent circuits.

2.3.1   The transponder antenna alone, in the "idle" state

We will begin by looking at the possible equivalent circuits of a transponder antenna when it is completely "idle", in other words, when it is not tuned, not loaded and not excited by an incident signal.

Natural or physical circuit The natural equivalent series circuit consists of the physical elements that form the antenna (see *Figure 2.1a*):

- a pure inductance due to the formation of a winding of wire or other substance, L_2s;

- a purely ohmic resistance of the winding, connected in series with L_2s, denoted R_2s;

- in parallel with everything else, a parasitic capacitance due to its construction technology, C_2p.

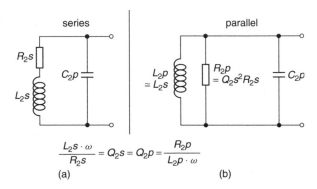

Figure 2.1   Natural circuit of the transponder antenna

The value of this capacitance is determined by the inter-turn capacitance, but also by the mechanical support on which the coil is created (a former, a printed circuit, plastic film, etc.). This system, namely, the parallel tuned circuit L_2s, C_2p, clearly has

- an intrinsic resonant frequency such that

$$L_2s \cdot C_2p \cdot \omega^2 = 1;$$

- an intrinsic quality factor, Q_2s, whose value clearly depends on the frequency at which the antenna will be operating:

$$Q_2s = \frac{(L_2s \cdot \omega)}{R_2s}.$$

Parallel equivalent circuit The natural circuit of the antenna can be depicted as a "parallel equivalent" circuit consisting of the following elements (see *Figure 2.1b* and see also the technical appendix at the end of this book):

$$L_2p = L_2s \frac{1 + Q_2s^2}{Q_2s^2} \text{ and, if the value of } Q_2s \text{ is large, } L_2p = L_2s;$$

$$R_2p = (Q_2s^2 + 1)R_2s \text{ and if the value of } Q_2s \text{ is large, } R_2p = R_2s \cdot Q_2s^2;$$

$$C_2p = C_2s \frac{(1 + Q_2s^2)}{Q_2s^2} \text{ and if the value of } Q_2s \text{ is large, } C_2p = C_2s;$$

and its intrinsic resonant frequency will also be such that $L_2p \cdot C_2p \cdot \omega^2 = 1$ (the same as before, naturally) and its quality factor can now be written:

$$Q_2p = \frac{R_2p}{L_2p \cdot \omega}$$

this value being the same as before, since

$$Q_2p = \frac{R_2p}{L_2p \cdot \omega} = \frac{Q_2s^2 \cdot R_2s}{L_2p \cdot \omega} = \frac{\dfrac{(L_2s \cdot \omega)^2}{R_2s^2} R_2s}{L_2p \cdot \omega}$$

and, since $L_2p = L_2s$

$$Q_2p = Q_2s$$

To simplify the explanation, in a first approximation, we may generally assume that the value of the capacitance C_2p is zero, and that the antenna winding can be represented simply by its inductance and its resistance.

2.3.2 Transponder antenna alone, "in the presence of a magnetic field"

The magnetic field $H(d, r)$, the induction $B(d, r)$ and the associated flux $\Phi(d, r)$ surrounding the transponder antenna cause the appearance, within the "idle" equivalent circuit, of a generator of an induced voltage V_{ind}, as shown in *Figure 2.2*. What is its value? Well, that is the question!

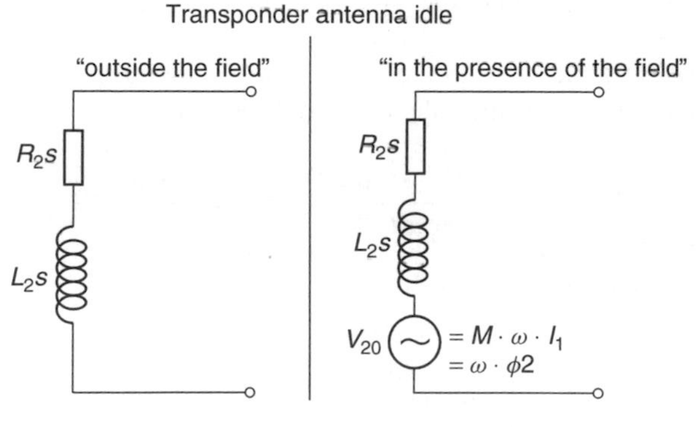

Figure 2.2 Transponder antenna idle, in the presence of a magnetic field

Value of the induced voltage across the terminals of the transponder antenna in the idle state In general, the value of the "idle" induced voltage V_{20} – of 2 in the secondary, and 0 in the idle state – (see *Figure 2.3*) developed across the terminals of the N_2 windings of the transponder coil as a result of a variation of flux $\Phi(d, r)$ as a function of the time (t), called $\Phi_2(t)$ for this purpose, is as follows:

$$V_{20}(t) = -\frac{d\Phi_2(t)}{dt} = -\frac{d}{dt}\int_{S_2} B \cdot dS_2 = -\mu \int_{S_2} H \cdot dS_2$$

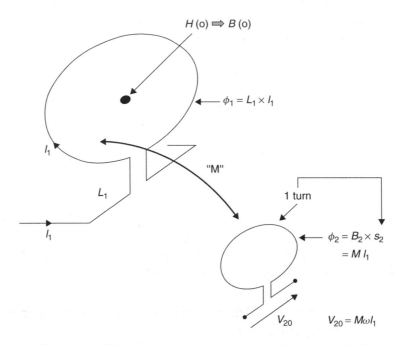

Figure 2.3   Voltage induced in the presence of a magnetic field

The sign "-" indicates the polarity of the induced voltage with respect to the phase of the magnetic flux that creates it.

If the variation of the flux $\Phi_2(t)$ is sinusoidal, its equation can be written as $\Phi_2(t) = \Phi_2 \cdot \sin(\omega t)$, where Φ_2 represents the rms value of the flux received by the totality of the turns ($\Phi_2 = N_2 \cdot B_2 \cdot s_2$).

This voltage $V_{20}(t)$ induced in the idle state across the terminals of the antenna coil (the antenna winding being in the idle state) will be equal to

$$V_{20}(t) = \frac{\mathrm{d}(\Phi_2 \cdot \sin(\omega t))}{dt} = -\Phi_2 \frac{\mathrm{d}\sin(\omega t)}{dt}$$

that is, after derivation

$$V_{20}(t) = -\omega \cdot \Phi_2 \cdot \cos(\omega t)$$

The effective value V_{20} of the induced voltage across the terminals (in the idle state) of the transponder antenna coil will therefore be

$$\boxed{V_{20}\mathrm{rms} = \omega \Phi_2}$$

or alternatively

$$\boxed{V_{20}\mathrm{rms} = \omega[B_2(N_2 \cdot s_2)]}$$

For a given value of the radius r of the base station, where H_d is the magnetic field strength at the distance d and allowing for the relation $B_2 = \mu \cdot H_d$, we find

$$V_{20}\text{rms} = \omega(\mu \cdot H_d)(N_2 \cdot s_2)$$

NOTE

The two equations above clearly show that the induced voltage V_{20}rms depends directly on ω, that is, on the frequency f. Additionally, as will be pointed out later in this book, the frequencies of 125 kHz and 13.56 MHz are commonly encountered among the frequencies used in contactless technology. Questions are often raised concerning the differences between the antenna technologies of transponders using these two frequencies.

Leaving aside all controversy and petty arguments, we can say, technically speaking, that

- in order to be able to use the same integrated circuit (or the same integrated circuit technology), in other words, one that requires the same induced voltage V_{20}rms to operate correctly,

- and assuming that the available effective physical surface area s_2 of the transponder is the same in both cases of use (e.g. for access control, regardless of the frequency used, and the use of a conventional smart card format),

- and, finally, regardless of the frequency, that the base station produces the same induction B_2 at the same distance,

then the preceding equations indicate that, in order to obtain the same voltage V_{20}rms we must keep the product ωN_2 constant.

For a simple picture of this, take the example of frequencies close to those used currently.

If, for example, $f = 130$ kHz and $f = 13$ MHz, giving a ratio of 100, as indicated in the lines above, then if everything else is equal, if we wish to obtain the same voltage V_{20}rms we need 100 times more turns N_2 at 130 kHz than at 13 MHz.

In principle, there is no reason not to provide these. However, this can cause serious technical problems, since, as we shall see, at 13.56 MHz we usually need only a few turns (3 to 5) that can easily be produced by printed circuit technology, whereas at 125 kHz (needing 300 to 500 turns) this is practically impossible.

On the other hand, if we consider the flux applied and introduce the value of the mutual inductance M present between the base station antenna and the transponder antenna, we can also write

$$\Phi_2(t) = M \cdot I_1(t)$$

$I_1(t)$ being the current flowing through the base station antenna.
Continuing in the same way, we find

$$V_{20}\text{rms} = \omega \cdot \Phi_2 = \omega \cdot M \cdot I_1$$

You should bear in mind that this voltage V_{20}rms is the voltage that will be measured, for example, in the measurement of the coupling coefficient k described in Chapter 10.

NOTE

In the literature, these equations are written in many different forms. Given that the signals in question are sinusoidal, the equations can also be written in complex form; for example, if

$$y(t) = Y \cdot \sin(\omega t + \varphi) = Y \cdot e^{(j\omega t)}$$

because

$$e^{(j\omega t)} = \sin(\omega t) + j\cos(\omega t)$$

then

$$\frac{d(y(t))}{dt} = Y \cdot j\omega e^{(j\omega t)} = j\omega \cdot Y(t)$$

If the variation of the flux, $\Phi_2(t)$, is sinusoidal, then if it is written with the use of the mutual inductance M, its equation will be

$$\Phi_2(t) = M \cdot I_1(t) = M(I_1 \sin(\omega t + \varphi))$$

$$= M(I_1 \cdot e^{(j\omega t)})$$

$$V_{20}(t) = -\frac{d\Phi_2(t)}{dt} = -M \cdot I_1 \cdot j\omega e^{j\omega t}$$

$$= -(j\omega \cdot M \cdot I_1 \cdot \sin(\omega t + \varphi))$$

$$= -j\omega \cdot M \cdot I_1(t)$$

and, therefore, as an effective value

$$V_{20}\text{rms} = \omega \cdot M \cdot I_1 = \omega \cdot \Phi_2$$

Q.E.D.

Value of the mutual inductance   Combining the above equations, we can write

$$V_{20}\text{rms} = \omega \cdot \mu \cdot H_d \cdot N_2 \cdot s_2 = M \cdot \omega \cdot I_1$$

and, replacing H_d with its value, we obtain

$$\omega \cdot \mu \left(\frac{r^2}{2(r^2 + d^2)^{3/2}} N_1 \cdot I_1 \right) N_2 \cdot s_2 = M \cdot \omega \cdot I_1$$

From this equation we can deduce the value of the mutual inductance M, that is,

$$M = \mu \frac{r^2}{2(r^2 + d^2)^{3/2}} N_1 \cdot N_2 \cdot s_2$$

Value of the coupling coefficient and operating range   Since we also know (see the proof in Chapter 10) that $M = k\sqrt{L_1 \cdot L_2}$, we also find that

$$\mu \frac{r^2}{2(r^2 + d^2)^{3/2}} N_1 \cdot N_2 \cdot s_2 = k\sqrt{L_1 \cdot L_2}$$

This equation is very useful, because, for a desired value of $V_{20}\text{rms}$, we can derive two fundamental pieces of information, namely, the coupling coefficient and the maximum operating range.

Coupling coefficient   The theoretical value of the coupling coefficient k as a function of the mechanical, electrical and physical parameters of the application is

$$k = \left(\mu \frac{r^2}{2(r^2 + d^2)^{3/2}} \right) N_1 \cdot N_2 \cdot s_2 \frac{1}{\sqrt{L_1 \cdot L_2}}$$

Everything else being equal, *Figure 2.4* shows the trend of the variations of k as a function of the distance d. Generally, the value of k is of the order of one per cent or less.

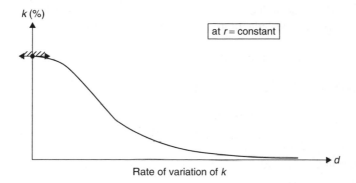

Figure 2.4   Variation of coupling coefficient k as a function of distance

Operating range   If the value of k is specified or known following a measurement made previously by a method such as that described in Chapter 10, we can estimate the value of the theoretical operating range d which can be achieved:

$$(r^2 + d^2)^{3/2} = \left(\mu_0 \frac{r^2}{2k} \right) N_1 \cdot N_2 \cdot s_2 \frac{1}{\sqrt{L_1 \cdot L_2}}$$

assuming that

$$A = \left(\mu_0 \frac{r^2}{2k} \right) N_1 \cdot N_2 \cdot s_2 \frac{1}{\sqrt{L_1 \cdot L_2}}$$

we find

$$d = \sqrt{A^{2/3} - r^2}$$

or we can estimate the value of the antenna radius required if we are to have, at a distance d, the desired coupling coefficient k.

AUTHOR'S NOTE

Don't be alarmed: these calculations are very simple, very quick and very informative, especially if they are done automatically in sequence on a good spreadsheet!

Concerning the magnetic flux   Let us re-examine *Figure 2.3*.
    The primary circuit produces a total flux, Φ_1, as follows:

$$\text{total flux} = \Phi_1 = L_1 \cdot I_1$$

The secondary circuit receives only a part of the total flux, known as the usable flux. This flux, Φ_2, can be expressed in various ways;

— on the one hand,
$$\text{usable.flux} = \Phi_2 = M \cdot I_1.$$

— and on the other hand,
$$\Phi_2 = N_2 \cdot \Phi_{21}$$

where Φ_{21} is the elementary flux per turn captured by the secondary, that is, $\Phi_{21} = B_2 \cdot s_2$, where B_2 is the induction present in the secondary (which will subsequently be termed B_d, that is, the induction at a distance d), and therefore,

$$\text{usable.flux} = \Phi_2 = N_2 \cdot B_2 \cdot s_2$$

From the above set of equations, we obtain

$$\frac{\text{usable.flux}}{\text{total flux}} = \frac{\Phi_2}{\Phi_1} = \frac{M}{L_1}$$

We will show in Chapter 10 that $M = k\sqrt{L_1 \cdot L_2}$, and therefore

$$\frac{\text{usable.flux}}{\text{total flux}} = k\frac{\sqrt{L_1 \cdot L_2}}{L_1}$$

Consequently,

$$\frac{\text{usable.flux}}{\text{total flux}} = k\sqrt{\frac{L_2}{L_1}}$$

Clearly, if $L_1 = L_2$, this ratio is equal to k!
    On the other hand, we have already shown that

$$V_{20} = \omega \cdot \Phi_2$$

and that the voltage across the terminals of the primary is:

$$V_1 = L_1 \cdot \omega \cdot I_1 = \omega \cdot \Phi_1$$

and therefore
$$\frac{\text{usable.flux}}{\text{total flux}} = \frac{\Phi_2}{\Phi_1} = \frac{V_{20}}{V_1}$$

Combining the last two equations, we obtain

$$\boxed{\frac{V_{20}}{V_1} = k\sqrt{\frac{L_2}{L_1}}}$$

an equation which will be found in quite a different way in Chapter 10.

2.3.3 Transponder antenna "tuned", but "unloaded", in the presence of a magnetic field

Now assume that the antenna coil of the transponder is tuned by providing a capacitance C in parallel with its winding (see *Figure 2.5a*).

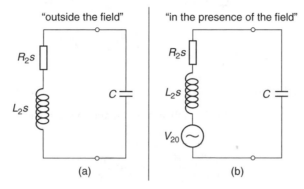

Transponder antenna tuned but unloaded

Figure 2.5   Transponder antenna tuned, but unloaded, in the presence of a magnetic field

If no magnetic field is present, the RLC tuned circuit set up in this way has an intrinsic resonant frequency and an intrinsic quality factor

$$Q_2s = \frac{L_2s \cdot \omega}{R_2s} = \frac{1}{R_2s \cdot C \cdot \omega}.$$

Now assume that the whole circuit is again placed in a magnetic field.

The presence of this field produces the same pd $V_{20}(t)$ as before in the antenna.

This means that the equivalent circuit must be modified to allow for this phenomenon, by connecting, in series with the inductance and resistance of the coil, a voltage generator with the value $V_{20}(t)$, as shown in *Figure 2.5b*.

Equivalent active dipole   Physically, in spite of the number of components shown in the equivalent circuit diagram, this circuit has only two terminals, A and B. Seen from these two terminals, this circuit can be considered to be an equivalent active dipole, consisting of an equivalent voltage generator $V_{20}e(t)$ and having an internal impedance of *Zie* (see *Figure 2.6*). To calculate these two values, let us work through the parallel and series transformations of the original circuit.

Value of the equivalent internal impedance of the antenna, Zie   The internal impedance *Zie* (complex value) of this active dipole consists of the impedance L_2s in series with R_2s, both in parallel with C (representing all capacitances combined in a single value, C). This means that

$$zie = \frac{(R_2s + j(L_2s \cdot \omega))\dfrac{1}{jC\omega}}{R_2s + j(L_2s \cdot \omega) + \dfrac{1}{jC\omega}}.$$

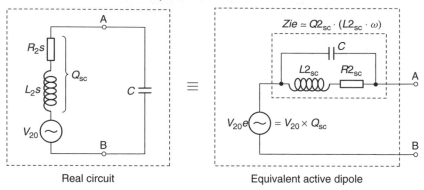

Figure 2.6   Active dipole equivalent to the transponder antenna alone

and therefore

$$zie = \frac{R_2s + j(L_2s \cdot \omega)}{(1 - L_2s \cdot C \cdot \omega^2) + jR_2s \cdot C \cdot \omega}$$

If, on the one hand, the transponder is tuned to the carrier frequency (f_c), and therefore $L_2s \cdot C \cdot \omega_c^2 = 1$ (which, for many reasons, is not always true or even desirable) we find that

$$zie = \frac{1 + \left(j\dfrac{L_2s}{R_2s} \right)\omega_c}{jC\omega_c}$$

and if, on the other hand, $Q_2s = (L_2s \cdot \omega_c)/R_2s$ is large (which is always true, with respect to 1), then Zie is given by

$$\boxed{zie = \frac{jQ_2s}{jC\omega_c} \rightarrow zie = Zie = Q_2s(L_2s \cdot \omega_c)}$$

Value of $V_{20}e$ To calculate the value of $V_{20}e$, we must first calculate the short-circuit current I_{cc} of the initial circuit. This is given by

$$I_{cc} = \frac{V_{20}}{R_2s + jL_2\omega_c}$$

and therefore

$$V_{20}e = I_{cc} \cdot zie$$

$$= \frac{V_{20}}{R_2s + jL_2s \cdot \omega_c}zie$$

$$= \frac{V_{20}}{jR_2s \cdot C\omega_c}$$

giving the value of the modulus of $V_{20}e$:

$$\boxed{V_{20}e = V_{20} \cdot Q_2 s}$$

Summary   We have thus defined the equivalent "active dipole" (see *Figure 2.7a*) representing the tuned antenna in the presence of a magnetic field but without loading by the integrated circuit. It takes the form of a *generator of an effective voltage of*

$$\boxed{\begin{array}{l} \text{ddp} \quad V_{20}e = Q_2 s \cdot V_{20} \\[1.5em] \text{internal impedance } zie = \dfrac{(R_2 s + \text{j}(L_2 s \cdot \omega))\dfrac{1}{\text{j}C\omega}}{R_2 s + \text{j}(L_2 s \cdot \omega) + \dfrac{1}{\text{j}C\omega}} \end{array}}$$

If the transponder is tuned to the carrier frequency f_c (the operating frequency ω_c is equal to the intrinsic frequency ω), and if $Q_2 s \gg 1$, the modulus Zie of the internal impedance is as follows:

$$Zie = Q_2 s(L_2 s \cdot \omega_c)$$

Figure 2.7b provides another equivalent "parallel" representation.

Now we simply need to load this equivalent dipole with the impedance R_{ic} represented by the integrated circuit of the transponder.

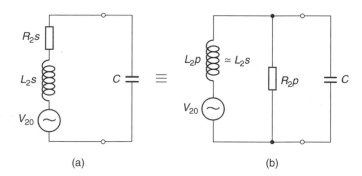

(a)   (b)

Figure 2.7   Equivalent active dipole: parallel representation

2.3.4   Transponder antenna "tuned and loaded" outside the magnetic field

The previous sections have described the equivalent parallel circuit of the transponder antenna when it is "tuned", but not connected to a load.

Now assume that the antenna is loaded with the input impedance of the integrated circuit of the transponder (generally R and C in parallel) (see *Figure 2.8*).

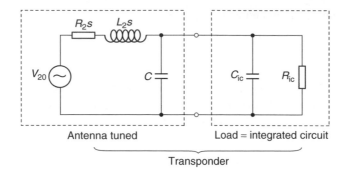

Antenna tuned   Load = integrated circuit

Transponder

Figure 2.8   Theoretical circuit of the complete transponder, outside the magnetic field (antenna loaded)

Complete equivalent transponder circuit   *Figure 2.9* shows the final equivalent circuit.

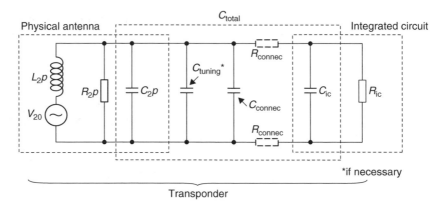

Transponder

Figure 2.9   Real circuit of the complete transponder, outside the field (antenna loaded)

After the usual "series/parallel" conversions, of the complete circuit consisting of the tuned antenna plus integrated circuit, in its parallel version, we can state that

$$L_2p = L_2s$$

$$Cp = C_{\text{tuning}} + C_{\text{ic}} + C_{\text{connec}} + C_2p$$

Cp = the sum of

- the tuning capacitance (if necessary);
- the input capacitance of the integrated circuit;
- the parallel capacitances of connections;
- the parallel capacitance C_2p of the antenna coil.

$$Rp = \frac{R_{\text{ic}} R_2 p}{R_{\text{ic}} + R_2 p}$$

Rp = the paralleling of the input resistance of the integrated circuit and the equivalent parallel resistance of the antenna coil.

And,

$$L_2 p \cdot Cp \cdot \omega^2 = 1$$

and

$$Qp_2 = \frac{Rp}{L_2 p \cdot \omega}$$

IMPORTANT NOTE
To avoid any subsequent confusion and headaches, I must give the reader due warning at this point about the similarity between the symbols Qp_2 and $Q_2 p$:
   Qp_2 above represents the quality factor of the whole assembly, consisting of the antenna and the integrated circuit;
   $Q_2 p$ simply represents the quality factor of the antenna winding only.

2.3.5 Transponder antenna "tuned and loaded" in the presence of a magnetic field

Now assume that the transponder in this state (with the antenna tuned and loaded by the integrated circuit) is placed in a magnetic field.

The presence of the variable magnetic field strength and induction, $H(t)$ and $B(t)$, and the frequency ω will, as explained above, create an induced voltage having the effective value V_{20} in the winding of the antenna coil. The new diagram for the equivalent circuit for the assembly, in the presence of a magnetic field, is shown in *Figure 2.10*. We shall now show that this assembly, consisting of the transponder antenna circuit and the integrated circuit, forms, on the one hand, an energy generation/load pair, and, on the other hand, a "voltage divider bridge".

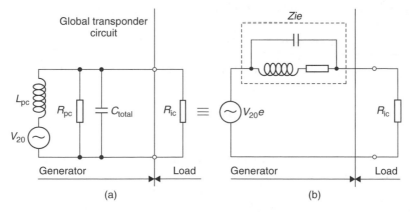

Figure 2.10   Transponder antenna, tuned and loaded, in the presence of a magnetic field

Energy generation/load pair   As indicated in the previous section, we are dealing with a pair consisting of an energy generator (the equivalent active dipole) and its load. The term "energy generator/load" pair implies a transfer of energy from the energy source to the load, and the aim of matching impedances and achieving optimal energy transfer.

Power matching of the transponder antenna If we try to optimize the system, in order to retain every last fraction of the energy supplied by the base station and thus provide long operating ranges, we must examine a new problem, namely, that of the best power matching between the energy generator represented by the antenna (via its induced voltage and its impedance) and the input impedance of the integrated circuit of the transponder.

Power matching condition for optimal power transfer In any transfer of power, the maximum value of the latter, called "power matching" (and consequently the maximum operating range, all other parameters being equal) is found when the real value of the load impedance, that is, the input impedance of the integrated circuit of the transponder, R_{ic}, is equal to the impedance of the power source (see *Figure 2.11*), in this case the impedance Zie of the antenna circuit.

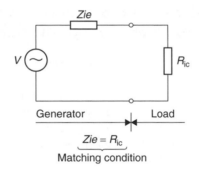

Figure 2.11 Impedance matching condition

To maximize power matching, it is therefore necessary that

$$Zie = R_{ic} \Rightarrow \tilde{R_{ic}} = (L_2s \cdot \omega)Q_2s$$

which, in this case, leads to the conclusion that the value of L_2s must be

$$L_2s = \frac{R_{ic}}{\omega \cdot Q_2s}$$

and, since $L_2p = L_2s$, there must then be a general parallel tuning capacitance, Cp, such that $L_2p \cdot Cp \cdot \omega = 1$.

NOTES

a) Clearly, for many understandable reasons (parasitic capacitances of connections, objectives or constraints relating to the detection of transponders arranged in stacks, collision management methods and procedures, etc.), these values are not exactly maintained, and the value of L_2s is therefore slightly adjusted to obtain the desired tuning.

b)   In this case, the pd V_{ic} across the terminals of the integrated circuit of the transponder will be

$$V_{ic} = \frac{R_{ic}}{R_{ic} + Zie} V_{20}e$$

where $R_{ic} = Zie$, and therefore

$$V_{ic} = \frac{1}{2} V_{20}e$$

or alternatively

$$V_{ic} = \frac{1}{2} Q_2 s \cdot V_{20}$$

We have not yet finished with the matter of the inductance of the transponder antenna, since, in order to achieve this impedance matching, we need to know R_{ic} and, unfortunately, this value is rather difficult to estimate.

Physical composition of the input impedance R_{ic} of the integrated circuit   Generally, the input circuit of the integrated circuit of the transponder consists of (see block diagram in *Figure 2.12*):

- a rectifier bridge (with conventional diodes or the equivalent using transistors);

- associated limiting circuits (e.g. voltage limiter, series or shunt regulator, etc.);

- strong field protection circuits, peak limiting devices;

- and so on.

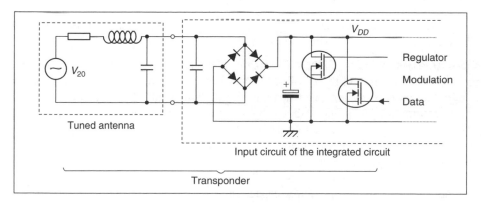

Figure 2.12   Physical composition of the input impedance of the integrated circuit

It is therefore fundamentally non-linear, meaning that everything changes as a function of everything else.

As will be shown in detail subsequently, this is particularly true of the values of the input capacitances and resistances as a function of the voltage V_{ic} present across the input terminals of the integrated circuit.

Let us assign some realistic orders of magnitude to the input impedance.

In the presence of very weak magnetic fields   When the magnetic field is very weak and the voltage V_{ic} is lower (of the order of 0.3 Vrms, for example) than the minimum operating threshold of the transponder, the latter receives no power (or insufficient power) and therefore does not operate; it has a very high impedance, and therefore certain values of input capacitance and resistance (generally of the order of several hundred kilohms, 100 to 200 kΩ, depending on the model and manufacturer), and thus, for a given inductance of the transponder antenna, there is also a certain intrinsic resonant frequency f_0 and a certain value of Q of the transponder.

In the presence of weak magnetic fields   When the magnetic fields are weak, in other words, just exceeding the minimum voltage threshold V_{ic} (with an order of magnitude from 2 to 4 Vrms, depending on the type and manufacturer of the integrated circuit), but not yet reaching the operating threshold of the voltage limiter or shunt regulator, the impedance Z_t is of the order of 10, 50 or 100 kΩ, resulting in new values for f_0' and Q'.

In the presence of stronger magnetic fields   When the fields are stronger (or when the transponder has moved towards the base station antenna in the same application), the internal series/shunt regulator comes progressively into action, to regulate the supply voltage Z_t of the transponder's integrated circuit. The impedance of the transponder decreases markedly, as does f_0'' and Q''

In response mode   When the base station supplies a pure carrier frequency and the transponder is in its response phase, then, because the load modulation principle is used, Z_t takes much lower values (of the order of a few hundred ohms – 150 or 200 Ω – or, depending on the model, a few kilohms, 2 to 5 for example), since this modulation operates strongly to enable the base station to understand the transponder. What do you think is happening?

f_0'' and Q''!

We will re-examine these matters more thoroughly in Chapter 6, in the description of the effect of strong magnetic fields on transponders.

Conclusions   Clearly, all these factors must be taken into account when dealing with the power matching and the possible variation of the resonant frequencies f_0 and the quality factors Q of the antenna when it is "idle", "half-loaded" or "loaded", and so on.

Concerning the last point, it should be noted that, to prevent an excessive mismatch $(f_0' - f_0)$ between the resonant frequency of the antenna circuit in the "idle" and "loaded" states, and to maintain approximately the same operating sensitivity at all times (reproducibility of the operating range), it can be shown that it is necessary to minimize the ratio $(f_0$ loaded$)/(f_0$ idle$)$, which depends both on the inductance of the transponder antenna and the absolute value of the tuning capacitance, as well as on its relative variations.

Consequently, each type of transponder (according to its size, winding technology, etc.) will have an optimal value of "idle" intrinsic frequency.

So we have made considerable progress: R_{ic} and C_{ic} depend on the voltage V_{ic} across the terminals of the integrated circuit – but what is the value of this voltage? The solution to this problem is explained below.

Value of induced voltage V_{ic}, with antenna loaded and tuned   Since the transponder antenna winding is no longer "idle" but "tuned" and "loaded" (in real-life operation) by the input impedance R_{ic} of the integrated circuit, on the basis of the redrawn equivalent circuit of the tuned antenna/integrated circuit assembly, now including the induced voltage generator V_{20}, we can calculate the voltage V_{ic} applied to the terminals of the integrated circuit ("divider bridge" circuit formed on the one hand by the combination of R_{ic} in parallel with Cp and on the other hand by $L_2s = L_2p$ in series with R_2s (see *Figure 2.13*)):

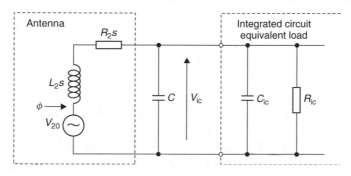

Figure 2.13   Voltage induced at the terminals of the integrated circuit

$$V_{ic} = \frac{R_{ic}//Cp}{Z_{L_{2p}} + (R_{ic}//Cp)} V_{20}$$

$$= \frac{\dfrac{R_{ic}}{1 + jR_{ic}Cp\omega}}{(R_2s + jL_2s\omega) + \dfrac{R_{ic}}{1 + jR_{ic}Cp\omega}} V_{20}$$

and therefore

$$V_{ic} = \frac{1}{1 + (R_2s + jL_2s\omega)\left(\dfrac{1}{R_{ic}} + jCp\omega\right)} V_{20}$$

As shown previously, in its complex representation the induced voltage V_{20} is as follows:

$$V_{20} = -j\omega \cdot M \cdot I_1$$

or alternatively

$$V_{20} = -j\omega \cdot B_d \cdot N_2 \cdot s_2$$

and therefore

$$V_{ic} = \frac{jM\omega I_1}{1 + (R_2s + jL_2s\omega)\left(\dfrac{1}{R_{ic}} + jCp\omega\right)} = \frac{-j\omega B_d \cdot N_2 \cdot s_2}{1 + (R_2s + jL_2s \cdot \omega)\left(\dfrac{1}{R_{ic}} + jCp\omega\right)}$$

in which equation $M = k\sqrt{L_1 L_2}$.

Let us re-examine the last equation and expand its denominator:

$$V_{ic} = \frac{-jM\omega I_1}{\left(1 - L_2 s \cdot Cp \cdot \omega^2 + \dfrac{R_2 s}{R_{ic}}\right) + j\left(\dfrac{L_2 s \cdot \omega}{R_{ic}} + R_2 s \cdot Cp \cdot \omega\right)}$$

We can deduce the real value of this equation, giving us

$$V_{ic}\frac{M\omega I_1}{\sqrt{\left(1 - L_2 s \cdot Cp \cdot \omega^2 + \dfrac{R_2 s}{R_{ic}}\right)^2 + \left(\dfrac{L_2 s \cdot \omega}{R_{ic}} + R_2 s \cdot Cp \cdot \omega\right)^2}} = f(\omega)$$

This equation, which is well known to RFID professionals, summarizes the reality of the mathematical relations between the various elements and provides the exact value of the voltage applied to the physical input of the integrated circuit. This is the value that is the most important "tool" for ensuring that a system operates correctly.

Figure 2.14 shows the variations of $V_{ic} = f(\omega)$.

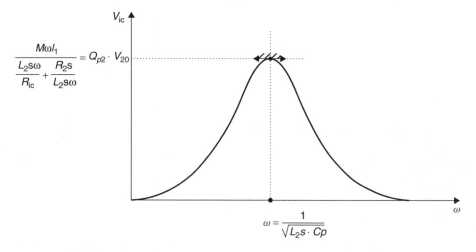

Figure 2.14   Variation of V_{ic} as a function of pulsation (i.e. frequency)

Tuned transponder ($\omega = \omega r$)   Now assume that the transponder is tuned to the exact frequency of the carrier supplied by the base station. In this case, $\omega_c^2 = \omega^2 = 1/(L_2 s \cdot Cp) \Rightarrow L_2 s \cdot Cp \cdot \omega^2 = 1$ and the term $1 - L_2 s \cdot Cp \cdot \omega^2 = 0$.

The general equation is then reduced to the following form:

$$V_{ic} = \frac{M\omega I_1}{\sqrt{\left(\dfrac{R_2 s}{R_{ic}}\right)^2 + \omega^2\left(\dfrac{L_2 s}{R_{ic}} + R_2 s \cdot Cp\right)^2}}$$

$$= \frac{M\omega I_1}{\sqrt{\left(\dfrac{R_2 s}{R_{ic}}\right)^2 + \dfrac{1}{L_2 s \cdot Cp}\left(\dfrac{L_2 s}{R_{ic}} + R_2 s \cdot Cp\right)^2}}$$

We now calculate the square of the denominator:

$$\left(\frac{R_2s}{R_{ic}}\right)^2 + \frac{1}{L_2s \cdot Cp}\left(\frac{L_2s}{R_{ic}} + R_2s \cdot Cp\right)^2 = \left(\frac{R_2s}{R_{ic}} \cdot \frac{R_2s}{R_{ic}+2}\right)$$
$$+ \left(\frac{L_2s}{Cp \cdot R_{ic}^2} + \frac{R_2s^2 \cdot Cp}{L_2s}\right)$$

If $R_2s \ll R_{ic}$, which is always the case, the first term of this equation is very small. We are left with

$$\frac{L_2s}{Cp \cdot R_{ic}^2} + \frac{R_2s^2 \cdot Cp}{L_2s}$$

In other words, also in the conditions below, the denominator will be practically equal to

$$\frac{1}{R_{ic}}\sqrt{\frac{L_2s}{Cp}} + R_2s\sqrt{\frac{Cp}{L_2s}}$$

Also, if we have $L_2s \cdot Cp \cdot \omega^2 = 1$, and therefore $Cp = 1/(L_2s \cdot \omega^2)$, the denominator will be

$$\frac{1}{R_{ic}}\sqrt{L_2s \cdot L_2s \cdot \omega^2} + R_2s\sqrt{\frac{1}{L_2s \cdot \omega^2 \cdot L_2s}}$$
$$= \frac{L_2s \cdot \omega}{R_{ic}} + \frac{R_2s}{L_2s \cdot \omega}$$

and therefore, in conclusion,

$$V_{ic}(\omega = \omega r) = \frac{M\omega I_1}{\sqrt{\left(\frac{R_2s}{R_{ic}}\right)^2 + \frac{1}{L_2s \cdot Cp}\left(\frac{L_2s}{R_{ic}} + R_2s \cdot Cp\right)^2}} = \frac{M\omega I_1}{\dfrac{L_2s \cdot \omega}{R_{ic}} + \dfrac{R_2s}{L_2s \cdot \omega}}$$

What is the physical meaning of the term $(L_2s \cdot \omega)/R_{ic} + R_2s/(L_2s \cdot \omega)$?
To discover this, we shall examine a circuit equivalent to the initial circuit, in which we have paralleled the series resistance of the antenna (see *Figure 2.15*). Assuming that

$$Q_2s = Q_{\text{antenna only}} = \frac{L_2s \cdot \omega}{R_2s}$$

$$R_2p = (Q_2s^2 + 1)R_2s$$

if Q_2s is large, we can state that $Q_2s^2 + 1 = Q_2s^2$, and therefore

$$R_2p = Q_2s^2 \cdot R_2s = \frac{(L_2s \cdot \omega)^2}{R_2s}$$

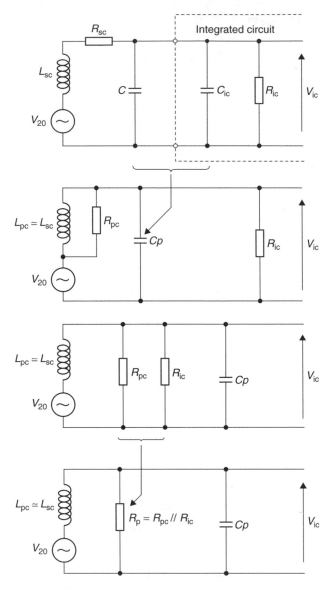

Figure 2.15   Damping of the transponder

Let us assume that

$$Rp = R_2p // R_{ic} = \frac{R_2p \cdot R_{ic}}{R_2p + R_{ic}}$$

and replace R_2p with its value in the preceding equation. We find

$$Rp = \frac{\dfrac{(L_2s \cdot \omega)^2}{R_2s} R_{ic}}{\dfrac{(L_2s \cdot \omega)^2}{R_2s} + R_{ic}}$$

$$Rp = \frac{R_{ic}\dfrac{(L_2s \cdot \omega)^2}{R_2s}}{\dfrac{(L_2s \cdot \omega)^2}{R_2s} + R_{ic}}$$

Let us now attempt to extract from this expression the ratio $(L_2s \cdot \omega)/Rp$. After expansion and reduction, we have

$$\frac{L_2s \cdot \omega}{Rp} = \frac{L_2s \cdot \omega}{R_{ic}} + \frac{R_2s}{L_2s \cdot \omega}$$

The equation given in the preceding section can therefore be written as

$$V_{ic} = \frac{M\omega I_1}{\dfrac{L_2s \cdot \omega}{R_{ic}} + \dfrac{R_2s}{L_2s \cdot \omega}} = \frac{M\omega I_1}{\dfrac{L_2s \cdot \omega}{Rp}}$$

Defining Qp_2 (remember not to confuse it with Q_2p!) as the value of the quality factor of the equivalent circuit formed by placing R_2p in parallel with R_{ic}, that is,

$$Qp_2 = \frac{Rp}{L_2s \cdot \omega}$$

we find that the ratio $(L_2s \cdot \omega)/Rp$ represents very exactly its reciprocal, that is,

$$\frac{L_2s \cdot \omega}{R_{ic}} + \frac{R_2s}{L_2s \cdot \omega} = \frac{1}{Qp_2}$$

$$= \frac{L_2s \cdot \omega}{Rp} = \text{total damping factor of transponder}$$

and therefore, in conclusion,

$$V_{ic} = Qp_2 \cdot M \cdot \omega \cdot I_1$$

$$V_{ic} = Qp_2 \cdot V_{20}$$

These last two equations show that, everything else being equal (and the matching being retained), V_{ic} is a function of Qp_2 and therefore of L_2s.

IMPORTANT NOTE

As we can see in these last equations, the voltage V_{ic} is much higher (by a factor of Qp_2) than the voltage V_{20} induced by the variation of flux. This is due to the fact that the antenna (and receiving) circuit of the transponder is tuned to (or very close to) the carrier frequency supplied by the base station. Consequently, the term "divider bridge" used for these explanations is, semantically, incorrect (please excuse me), but the topology used for the electrical description of the equivalent circuit requires the use of this phrase!

Value of induced voltage $V_{ic\,min}$, with antenna loaded and tuned   To enable a transponder integrated circuit to operate correctly and perform its functions, the voltage V_{ic} must reach at least a minimum value of $V_{ic\,min}$ specified by its manufacturer, even when the transponder is not tuned to the carrier frequency supplied by the base station! Let us re-examine the expression of its equation:

$$V_{ic} = \frac{M \cdot \omega c \cdot I_1}{\sqrt{\left(1 - L_2 s \cdot Cp \cdot \omega c^2 + \dfrac{R_2 s}{R_{ic}}\right)^2 + \left(\dfrac{L_2 s \cdot \omega c}{R_{ic}} + R_2 s \cdot Cp \cdot \omega c\right)^2}} = f(\omega c)$$

If the transponder is tuned to the carrier frequency ($\omega c = \omega r$, $L_2 s \cdot Cp \cdot \omega c = 1$), then according to what has been demonstrated in the previous sections, we can calculate the minimum value of the voltage $V_{20\,min}$ that must be developed in the idle state across the terminals of the transponder antenna. For this purpose, applying

$$V_{ic\,min} = \frac{Rp}{L_2 s \cdot \omega} \cdot V_{20\,min}$$

we find

$$V_{ic\,min} = Qp_2 \cdot V_{20\,min}$$

or alternatively

$$V_{20\,min} = V_{ic\,min} \cdot \frac{1}{Qp_2}$$

Let us re-examine the equation $V_{20} = M\omega I_1$, and replace the mutual inductance M with its value

$$M = k\sqrt{L_1 L_2}$$

giving us

$$V_{20} = k\sqrt{L_1 L_2} \cdot \omega I_1$$

By rearranging the parameters of this equation we can write it as follows:

$$V_{20} = k\sqrt{L_2} \cdot \frac{L_1 \omega I_1}{\sqrt{L_1}}$$

Now, the product $L_1 \omega I_1$ is equal to the voltage V_1 present across the terminals of the base station antenna. We therefore find that

$$V_{20} = k\sqrt{\frac{L_2}{L_1}} \cdot V_1$$

If the voltage V_1 is known or specified, then in order for the application to function correctly we can calculate k_{min} such that

$$V_{20\,min} = V_1 \cdot k_{min}\sqrt{\frac{L_2}{L_1}}$$

and therefore

$$k_{\text{min}} = \frac{V_{20\,\text{min}}}{V_1 \sqrt{\dfrac{L_2}{L_1}}}$$

Using the preceding equations, we obtain

$$V_{\text{ic min}} = V_{20\,\text{min}} \cdot Qp_2 = V_1 \cdot k_{\text{min}} \sqrt{\frac{L_2}{L_1}} \cdot Qp_2$$

and therefore

$$k_{\text{min}} = \frac{V_{\text{ic min}}}{V_1 Qp_2} \sqrt{\frac{L_1}{L_2}}$$

Let us bring this value back into the mutual inductance equation:

$$M_{\text{min}} = k_{\text{min}} \sqrt{L_1 L_2}$$
$$= \frac{V_{\text{ic min}}}{V_1 Qp_2} L_1$$

When $V_1 = L_1 \omega I_1$, we obtain

$$M_{\text{min}} = \frac{V_{\text{ic min}}}{Qp_2 \omega I_1}$$

$$M_{\text{min}} \omega I_1 = \frac{V_{\text{ic min}}}{Qp_2} = V_{20\,\text{min}}$$

Q.E.D.

These equations will be examined in more detail when we look at the downlink (communications from the transponder to the base station) in Chapter 6.

Maximum value of Qp₂   To find out whether $Qp_2 = f(L_2 s)$ passes through a maximum, let us examine the value and variations of the derivative $(d\,Qp_2)/(d\,L_2 s)$ of the function $Qp_2 = f(L_2 s)$

$$Qp_2 = \frac{1}{\dfrac{L_2 s \cdot \omega}{R_{\text{ic}}} + \dfrac{R_2 s}{L_2 s \cdot \omega}} = f(L_2 s)$$
$$= \frac{R_{\text{ic}} \cdot L_2 s \cdot \omega}{L_2 s^2 \cdot \omega^2 + R_2 s \cdot R_{\text{ic}}} = \frac{u}{v}$$

Let us examine the variations of $Qp_2 = f(\omega)$. To do this, we must calculate its derivative $Qp_2{}'$, which takes the form

$$Qp_2{}' = \frac{u'v - v'u}{v^2}$$

that is,

$$Qp_2' = \frac{R_{ic}L_2s(R_2s \cdot R_{ic} - L_2s^2 \cdot \omega^2)}{v^2}$$

This derivative Qp_2' is cancelled for $L_2s \cdot \omega = \sqrt{R_2s \cdot R_{ic}}$, that is,

$$L_2s = \frac{\sqrt{R_2s \cdot R_{ic}}}{\omega}$$

When the value of L_2s is equal to the value shown above, the function $Qp_2 = f(L_2s)$ passes through a maximum

$$\boxed{Qp_{2\,max} = \frac{1}{2}\sqrt{\frac{R_{ic}}{R_2s}}}$$

This last equation can be used to optimize the energy transmission by maximizing the value of the voltage V_{ic}.

Parallel representation of the transponder Let us return to the parallel equivalent general circuit discussed in the preceding sections (see *Figure 2.16*) in which there appears only Rp.

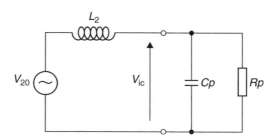

Figure 2.16 Parallel equivalent circuit of the transponder

The general equation for the voltage V_{ic} across the terminals of the integrated circuit is then written:

$$V_{ic} = \frac{Rp//Cp}{Z_{L_2p} + (Rp//Cp)}V_{20}$$

$$= \frac{\dfrac{Rp}{1 + jRpCp \cdot \omega}}{jL_2p \cdot \omega + \dfrac{Rp}{1 + jRpCp \cdot \omega}}V_{20}$$

After expansion and reduction, we have

$$V_{ic} = \frac{1}{(1 - L_2pCp \cdot \omega^2) + \dfrac{jL_2p \cdot \omega}{Rp}}V_{20}$$

Replacing V_{20} with its value, we obtain

$$V_{ic} = \frac{-jM\omega I_1}{(1 - L_2 p C p \cdot \omega^2) + j\dfrac{L_2 p \cdot \omega}{Rp}}$$

Using the value of the complex variable represented by the result of this equation, and assuming that $L_2 p C p \cdot \omega^2 r = 1$, we find the value of the modulus of V_{ic}:

$$V_{ic} = \frac{1}{\sqrt{\left[\left(1 - \left(\dfrac{\omega}{\omega r}\right)^2\right)^2 + \left(\dfrac{L_2 p \cdot \omega}{Rp}\right)^2\right]}} M\omega I_1$$

Everything else being equal, V_{ic} is proportional to I_1.

Given that, once again, $M I_1 = N_2 \Phi_{21} = B_d N_2 s_2 = \mu H_d N_2 s_2$, we obtain

$$V_{ic} = \frac{1}{\sqrt{\left[\left(1 - \left(\dfrac{\omega}{\omega r}\right)^2\right)^2 + \left(\dfrac{L_2 p \cdot \omega}{Rp}\right)^2\right]}} \omega \cdot \mu H_d N_2 s_2$$

Everything else being equal, V_{ic} is proportional to H_d.

Minimum threshold magnetic field This last equation can be used to deduce the relationship between the value of the magnetic field H_d (therefore for a distance d) as a function of the voltage V_{ic} (or the reciprocal, if you prefer), that is,

$$H_d = \frac{\sqrt{\left[\left(1 - \left(\dfrac{\omega}{\omega r}\right)^2\right)^2 + \left(\dfrac{L_2 p \cdot \omega}{Rp}\right)^2\right]}}{\omega \cdot \mu N_2 s_2} V_{ic}$$

To enable a transponder integrated circuit to operate correctly and perform its functions, the voltage V_{ic} must reach at least the minimum voltage, $V_{ic\,min}$ even if the transponder is not tuned to the carrier frequency sent from the base station. This will be reached for a value of the magnetic field called the threshold value, H_{dt} (dt as the threshold at a distance d) of

$$H_{dt} = \frac{\sqrt{\left[\left(1 - \left(\dfrac{\omega}{\omega r}\right)^2\right)^2 + \left(\dfrac{L_2 p \cdot \omega}{Rp}\right)^2\right]}}{\omega \cdot \mu N_2 s_2} V_{ic\,min}$$

or alternatively, by introducing Qp_2 into the equation

$$H_{dt} = \frac{\sqrt{\left[\left(1 - \left(\frac{\omega}{\omega r}\right)^2\right)^2 + \left(\frac{1}{Qp_2}\right)^2\right]}}{\omega \cdot \mu N_2 s_2} V_{ic\,min}$$

The curve of the example shown in *Figure 2.17* indicates the trend of the variations of H_{dt} as a function of ω (or of f) for a transponder integrated circuit whose minimum voltage $V_{ic\,min}$ would be specified.

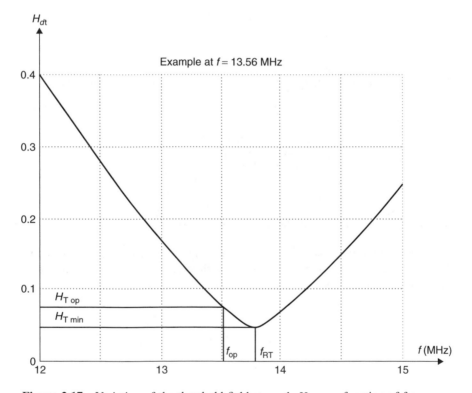

Figure 2.17   Variation of the threshold field strength H_{dt} as a function of frequency

Case of a tuned transponder   In the case in which the transponder is tuned to the carrier frequency, $\omega c = \omega r$.

In this case, if we return to the equations

$$V_{ic\,min} = V_{20\,min} Qp_2 = V_1 \cdot k_{min} \sqrt{\frac{L_2}{L_1}} \cdot Qp_2$$

$$= \frac{Rp}{L_2 p} \mu H_d N_2 s_2$$

and since $L_2p = L_2s$ and $Qp_2 = Rp/(L_2s \cdot \omega)$, that is, $Rp/(L_2p) = Qp_2 \cdot \omega$, the equation takes the simple form

$$B_d = \mu H_d = \frac{1}{Qp_2 \cdot \omega N_2 s_2} V_{ic}$$

or alternatively,

$$H_d = \frac{1}{\mu Qp_2 N_2 s_2 \omega} V_{ic}$$

giving, once again, the value of the threshold magnetic field:

$$H_{dt} = \frac{1}{\mu N_2 s_2} \cdot \frac{1}{Qp_2} \cdot \frac{1}{\omega} V_{ic\,min}$$

The last expression of this equation, on the one hand at given values of ω, N_2, s_2 and on the other hand for a desired operating voltage (for example $V_{ic\,min}$) can be used to trace the variations of $H_{dt} = f(Qp_2)$. A specific example of the variations of H_{dt} is shown in *Figure 2.18*.

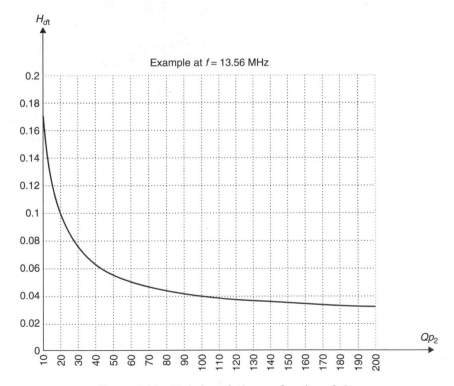

Figure 2.18   Variation of H_{dt} as a function of Qp_2

It should be noted in this example (parametered for 13.56 MHz) that, owing to the hyperbolic form of the equation $H_{dt} = f(Qp_2)$, it is rather pointless to try to increase at all costs the quality factor Qp_2 with the aim of greatly reducing the value of H_{dt}.

Now at last we know what we must do to maintain and guarantee the minimum value of V_{ic}, and therefore to be able to discover, for this specific voltage, the values of R_{ic} and C_{ic} of the integrated circuit, and therefore $Q_2 s$.

To put it another way, we can finally calculate $L_2 s - L_2 s = R_{ic}/(\omega Q_2 s)$ – to ensure the best power matching condition between the base station and the integrated circuit via the transponder antenna.

It is now up to you to define the voltage V_{ic} (or a range of voltages V_{ic}') at which you will guarantee the correct operation of the system.

Global impedance of the transponder circuit   Let us return to the equivalent circuit of the transponder (see *Figure 2.19*) and examine the value, without approximation, of the impedance represented by all of its components, seen from the generator terminals, as a function of the frequency at which the base station operates.

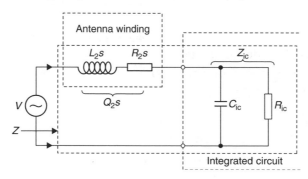

Figure 2.19   Global impedance of the transponder circuit

The expression of the complex impedance of the whole transponder circuit is as follows:

$$z = (L_2 s + R_2 s) \text{ in series with } (Cp \text{ and } R_{ic} \text{ in parallel})$$

$$= (R_2 s + j L_2 s \cdot \omega) + \frac{R_{ic}}{1 + j R_{ic} Cp \cdot \omega}$$

$$= \frac{[R_2 s + R_{ic}(1 - L_2 s Cp \cdot \omega^2)] + j\omega(L_2 s + R_2 s R_{ic} Cp)}{1 + j R_{ic} Cp \cdot \omega}$$

The modulus Z of the impedance is therefore

$$Z = \sqrt{\frac{[R_2 s + R_{ic}(1 - L_2 s Cp \cdot \omega^2)]^2 + (L_2 s + R_2 s R_{ic} Cp)^2 \omega^2}{1 + R_{ic}^2 Cp^2 \cdot \omega^2}}$$

Given the value of Z, we can easily deduce the effective current – i_2 – flowing in the transponder circuit, since

$$\frac{V_{20}}{Z} = i_2$$

or alternatively

$$\frac{\dfrac{V_{ic}}{Q_2}}{Z} = i_2$$

NOTE

In the case in which the transponder is tuned precisely to the carrier frequency ($L_2sCp\omega^2 = 1$), the exact equation for the modulus Z of the impedance is simplified, giving us

$$Z = \sqrt{\frac{R_2s^2 + (L_2s + R_2s\,R_{ic}Cp)^2\omega^2}{1 + R_{ic}{}^2Cp^2\omega^2}} = \sqrt{\frac{R_2s^2 + (L_2s + R_2s\,R_{ic}Cp)^2\omega^2}{1 + \dfrac{R_{ic}{}^2Cp}{L_2s}}}$$

An approximation-based method – but very accurate!   Very often it is easier to calculate the impedance of the whole transponder by making a number of approximations, which are, in fact, always borne out by the reality of applications.

Let us examine *Figure 2.20 (see next page)*, which shows the original equivalent circuit again, and let us now move from parallel to series with respect to the pair formed by the tuning capacitance and the resistance R_{ic}.

The equivalent value of the series resistance Rs_{ic} is then

$$Rs_{ic} = \frac{R_{ic}}{Q^2 + 1}$$

where $Q = R_{ic}Cp \cdot \omega$ and $Q \gg 1$.
Therefore,

$$Rs_{ic} = \frac{R_{ic}}{Q^2}$$

$$= \frac{1}{R_{ic}Cp^2\omega^2}$$

If the transponder is tuned to the carrier frequency (and therefore to the resonance) sent by the base station, we can write

$$1 = L_2sCp\omega^2 \Rightarrow 1^2 = (L_2sCp\omega^2)^2$$

and therefore

$$Rs_{ic} = \frac{(L_2sCp\omega^2)^2}{R_{ic}Cp^2\omega^2} = \frac{L_2s^2\omega^2}{R_{ic}}$$

and therefore, finally, the total series resistance of the circuit:

$$\boxed{Rs_{tot} = R_2s + \frac{L_2s^2\omega^2}{R_{ic}}}$$

Clearly, once again, by dividing the induced voltage V_{20} by the value of Rs_{tot}, we obtain the value of the current flowing through the circuit in resonance.

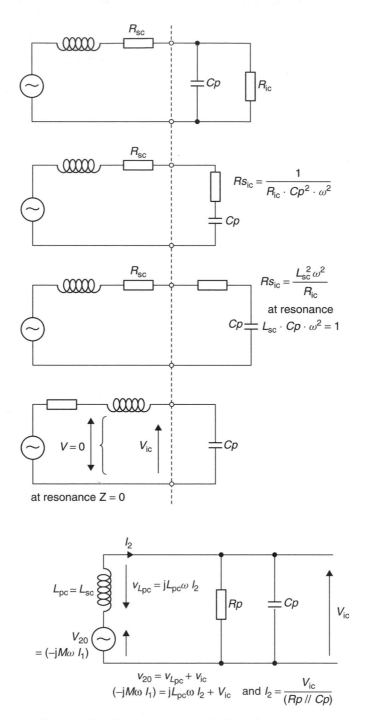

Figure 2.20   Series equivalent circuit of the transponder

This very simple and "friendly" equation is a demonstration and a reminder of several things:

If, $Rs_{tot} = R_2s + "R_2s" = 2\ R_2s$, then the matching between the generator (serialization impedance L_2s and R_2s) and the load (the integrated circuit consisting of the capacitance Cp and R_{ic} in parallel) is perfect, and therefore

$$R_2s = \frac{L_2s^2\omega^2}{R_{ic}}$$

As pointed out above, this, apparently by chance, is the condition for obtaining the maximum and optimal value of Qp_2! Isn't it strange how good things happen by chance...

NOTE

In the examples in Chapter 6, we shall see that, in spite of the approximations that we have made, the differences in practice are virtually non-existent.

2.3.6 Concluding remarks

We know that the induced voltage V_{20} is proportional to the number of turns, N_2, making up the coil of the transponder antenna.

Assuming that V_{20o} is the induced voltage that would be produced across the terminals of a single turn of the transponder antenna, we can write that $V_{20e} = N_2 V_{20o}$ and, allowing for the fact that $Q_2 = (L_2s \cdot \omega)/R_2s$ if there is perfect power matching, the equation takes the form

$$V_{ic} = \frac{1}{2}N_2 V_{20o}\frac{L_2s \cdot \omega}{R_2s}$$

For any given antenna coil technology (with a particular circular, square, flat or other shape, dimensions, arrangement of winding wires, etc.), the values of L_2s and R_2s also depend on N_2 and therefore the equation for V_{ic} is a complex function of $N_2 - V_{ic}(N_2)$ – through the values of N_2, $L_2s(N_2)$ and $R_2s(N_2)$.

As for the optimal power transfer, this will be present when the ratio V_{20}/V_{20o} is largest, meaning that we must calculate the value of N_2 at which this function passes through its maximum.

To do this, we must find the exact numerical form of the function $(V_{ic}(N_2))/V_{20o}$ and find the value of N_2 that cancels its first derivative.

These calculations cannot be shown here, as they essentially depend on the technology used to form the antenna coil, and can therefore only be carried out by the designer. Everyone has his own problems!

To illustrate these concepts, and purely by way of guidance, *Figure 2.21 (see next page)* shows an example of the variation of V_{ic}/V_{20o} as a function of N_2 in the case of a flat circular coil with very specific dimensions.

It should be noted, however, that at this stage all the principal parameters of the electrical components of the transponder and the number of turns of the transponder antenna coil have been determined.

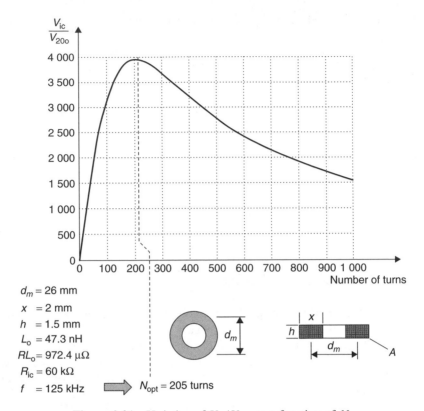

Figure 2.21 Variation of V_{ic}/V_{20o} as a function of N_2

2.3.7 Analysis of the transient response of the transponder to the signals from the base station

Binary data sent from the base station to the transponder are generally transmitted by a constant frequency carrier wave whose amplitude is modulated by the binary stream according to an "all or nothing" principle called amplitude shift keying (ASK), at 100% or 10% according to circumstances (see *Figures* 2.22a and 2.22b).

This total or partial carrier cut-off indicating the presence of a data element related to the value of the transmitted bit must be understood by the transponder before its subsequent interpretation. For this purpose, the voltage actually present across the terminals of the transponder integrated circuit must have a variation representing the incident signal, with an absolute value that can be used for its demodulation within the transponder integrated circuit.

As we have seen previously, in order to have more signal and therefore the greatest possible operating range, the transponder antenna circuit is tuned.

The tuned circuit formed in this way then has a global quality factor Qp_2 which should, theoretically, be as high as possible to obtain the best possible performance. However, if we carried this approach to extremes, making the value Qp_2 infinite, then, in spite of the variations of amplitude due to the ASK modulation of the base station carrier, there would be no variation of voltage across the terminals of the

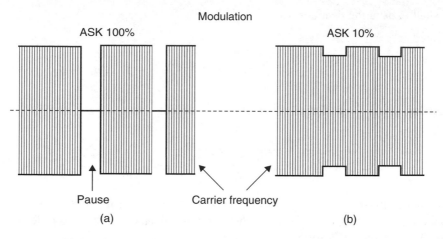

Figure 2.22   100% ASK and 10% ASK amplitude modulation

transponder integrated circuit. There is therefore a maximum optimal value of the Qp_2 of the transponder that must be observed. This value is therefore directly related to the transient response of the tuned circuit formed by the whole electrical circuit of the transponder.

Let us return to the equivalent circuit of the transponder (see *Figure 2.23*).

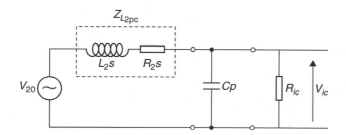

Figure 2.23   Global electrical circuit of the transponder

The equation for the voltage across the terminals of the integrated circuit:

$$V_{ic} = \frac{R_{ic}//Cp}{Z_{L_2p} + (R_{ic}//Cp)} V_{20}$$

and therefore

$$\frac{V_{ic}}{V_{20}} = \frac{\dfrac{R_{ic}}{1 + jR_{ic}Cp \cdot \omega}}{(R_2s + jL_2s \cdot \omega) + \dfrac{R_{ic}}{1 + jR_{ic}Cp \cdot \omega}}$$

that is,

$$\frac{V_{ic}}{V_{20}} = \frac{1}{1 + (R_2s + jL_2s \cdot \omega)\left(\dfrac{1}{R_{ic}} + jCp \cdot \omega\right)}$$

We shall expand the denominator:

$$\frac{V_{ic}}{V_{20}} = \frac{1}{1 + \dfrac{R_2 s}{R_{ic}} + jR_2 sCp \cdot \omega + \dfrac{jL_2 s \cdot \omega}{R_{ic}} - L_2 sCp \cdot \omega^2}$$

Assuming that $p = j\omega$, that is, $p^2 = -\omega^2$, we obtain

$$\frac{V_{ic}}{V_{20}} = \frac{1}{1 + \dfrac{R_2 s}{R_{ic}}} \cdot \frac{1}{1 + \dfrac{R_{ic} R_2 sCp + L_2 s}{R_{ic} + R_2 s} \cdot p + \dfrac{R_{ic} L_2 sCp}{R_{ic} + R_2 s} \cdot p^2}$$

Presented in this way, this equation corresponds to a special form of the established equation for the pulsed response to the transient signals of a second-order circuit, generally written in this form:

$$G(s) = \text{const} \; \frac{1}{1 + \dfrac{2\Delta}{\omega_0^2} p + \dfrac{1}{\omega_0^2} p^2}$$

in which equation *const* is a constant.

Taking the last two equations, we can easily identify, term by term, the values of ω_0^2 and 2Δ. Thus we find that

$$\omega_0^2 = \frac{R_{ic} + R_2 s}{R_{ic} L_2 sCp}$$

and

$$2\Delta = \frac{R_{ic} R_2 sCp + L_2 s}{R_{ic} L_2 sCp}$$

Assuming that

$$\theta = \frac{1}{\Delta} = \text{damping coefficient}$$

$$T_0 = \frac{1}{f_0}$$

$$1 = (L_2 sCp)\omega r^2 \Rightarrow \frac{1}{\omega r^2} = L_2 sCp$$

$$1 = (L_2 sCp)\omega r^2 \Rightarrow Cp \cdot \omega r = \frac{1}{L_2 s \cdot \omega r}$$

$$\omega r = 2\pi f_0 \Rightarrow 2 = \frac{\omega r}{\pi f_0}$$

$$\omega r = 2\pi f_0 \Rightarrow 2 = \frac{\omega r T_0}{\pi}$$

we can write

$$\theta = 2\frac{R_{ic}L_2sCp}{R_{ic}R_2sCp + L_2s}$$

$$= T_0\frac{\omega r}{\pi} \cdot \frac{\dfrac{1}{\omega r^2}}{R_2sCp + \dfrac{L_2s}{R_{ic}}}$$

$$= \frac{T_0}{\pi} \cdot \frac{1}{R_2sCp \cdot \omega r + \dfrac{L_2s \cdot \omega r}{R_{ic}}}$$

Given that $L_2sCp \cdot \omega r^2 = 1 \Rightarrow Cp \cdot \omega r = 1/(L_2s \cdot \omega r)$, we obtain

$$\theta = \frac{T_0}{\pi} \cdot \frac{1}{\dfrac{R_2s}{L_2s \cdot \omega r} + \dfrac{L_2s \cdot \omega r}{R_{ic}}}$$

Now, we have shown previously that $R_2s/(L_2s \cdot \omega r) + (L_2s \cdot \omega r)/R_{ic} = 1/Qp_2$, in which equation Qp_2 represents the global quality factor of the set of components making up the transponder (antenna, tuning capacitance, integrated circuit), and therefore, finally,

$$\boxed{\theta = \frac{T_0}{\pi}Qp_2 \Rightarrow \frac{\theta}{T_0} = \frac{Qp_2}{\pi}}$$

This shows that, in order to have a short transient time θ, its value must be small with respect to the period T_0 of the carrier frequency.

In this case, we often use relative values – the ratio θ/T_0 – rather than just the value of θ. On the other hand, the equation for θ clearly demonstrates its linear relationship with the global value of the quality factor Qp_2. The maximum value of Qp_2 – $Qp_{2\,max}$ – will therefore be related to the authorized value of θ/T_0 (which itself is related to the minimum demodulation threshold of the integrated circuit) so that the integrated circuit can "understand" the incident signal sent from the base station (see *Figure 2.24*).

It should also be noted that the value of Qp_2 ($Qp_2 = (L_2\omega r)/Rp$, where $Rp = R_2p//R_{ic}$) is closely dependent on the value of R_{ic} which, as emphasized before, is far from being linear, and whose value changes greatly between weak fields and strong fields.

It should be noted that a similar analysis, complementary to this one, will be undertaken for the same reasons in the next chapter, relating to the important electrical parameters of the base station.

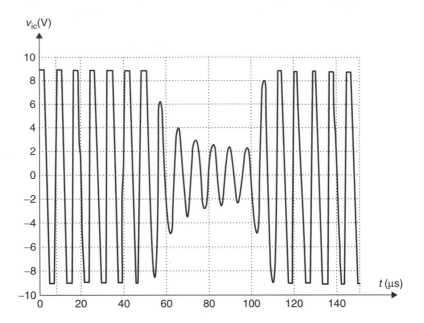

Figure 2.24   Example of transient voltage V_{ic} as a function of time

2.3.8   Reading transponders arranged in a stack

At various points in this chapter, I have mentioned that it is sometimes undesirable to tune the transponder to the carrier frequency sent from the base station. Having opened this subject several times, I shall now examine it more closely.

In many applications, the transponders have to be placed in a stack.

Figure 2.25 shows an example in which a number of contactless cards (public transport + office access cards + etc.) are stacked on each other or perhaps slightly fanned out in a case. We may find the same situation with letters in a mailbag or small packets arranged side by side on a pallet or books on a library shelf.

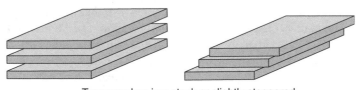

Transponders in a stack or slightly staggered

Figure 2.25   Transponders in a stack

The "off-centre tuning" method   The theory of coupled circuits, applied to the antennas of transponders arranged in a stack, indicates that all the antennas of the transponders interact with each other, causing initial and individual detuning, and making the tuning frequency of each antenna tend to decrease.

This is why, in such situations, it is useful to tune each transponder individually to a frequency approximately 25% to 40% higher than the carrier frequency: this is called the *off-centre tuning* method (e.g. in a system operating at 13.56 MHz, it is common practice to tune the transponders to approximately 1.25 fc = 16.95 MHz). In theory, all contactless cards in ISO format that may be stacked in a case are (or should be!) tuned to a frequency that is not exactly the same as the carrier frequency sent from the base station. Designers who do not allow for these circumstances in their applications are likely to encounter severe problems.

Clearly, this has repercussions on the well-known parameter of threshold field/magnetic induction (H_{dt}, B_{dt}) required for correct operation, and consequently on the operating range of each of the transponders considered individually and of the set of transponders when they are stacked.

A transponder produced according to the off-centre tuning method theoretically has a higher threshold magnetic field value than a transponder tuned to the carrier frequency (see *Figure 2.17*). For example, if everything else is equal, a contactless card or transponder tuned to 16.95 MHz can require approximately 1.2 A/m to operate correctly, as against 0.2 A/m for the same components tuned strictly to the carrier at 13.56 MHz.

Clearly, for the same energy provided by the base station, this method of off-centre tuning reduces the operating range, and therefore, seen from the outside, one transponder considered in isolation and a stack of transponders appear to have practically identical operating ranges. Magic, isn't it? Is there a trick? Of course!

When a number of transponders in a stack are simultaneously within the field of the base station antenna, the global tuning of all the transponders tends to return to the carrier frequency, and the threshold magnetic field value of each transponder decreases, which, theoretically, tends to increase the operating range; for practical purposes, this is incorrect (!) in the present case, since, at this instant, the reader supplies energy to all the transponders present in the field and its effective range is reduced at this particular moment, so that ultimately, as far as the users are concerned, an isolated transponder appears to operate at the same range as a stack of transponders (see also Chapter 6).

A last word. You will also have noticed that the magnetic field range for the operation of "proximity" contactless cards is of the order of 10 cm between the farthest distance and the nearest distance. This means, for example, that a transponder that, for its application, must be deliberately detuned to 16.95 MHz and requires 1.2 A/m to operate, must also withstand 12 A/m.

2.3.9 Examples

Many examples of applications (at 125 kHz and 13.56 MHz) are provided in Chapters 5 and 6 to illustrate all these calculations in detail.

3

The Base Station: Supplementary Information

Before examining the purely electronic part of the base station which is detailed in Chapter 9, and the signal processing aspects, we will begin by looking at what are generally considered to be the most complicated areas, namely, the theoretical, practical and technological definitions of the base station antenna.

3.1 The Base Station Antenna

In this chapter, we shall look at the main electrical parameters of the base station antenna. Chapter 8 will provide a detailed description of the different types of technology and implementations of antennae and their numerous individual and/or special characteristics, which are not dealt with in this chapter.

3.2 Review and Supplementary Technical Information

The following sections provide a brief review and supplementary information on the theory of RLC circuits, which we will require to design and optimize the parameters of base station antennae.

3.2.1 Maximum permissible quality factor of the base station antenna

As mentioned earlier in this book, most uplinks use ASK modulation of the carrier.

To ensure the correct operation of the binary part of the communication, the maximum value of the quality factor $Q_{1\,max}$ of the base station antenna must be such that the

RFID and Contactless Smart Card Applications   D. Paret
© 2005 John Wiley & Sons, Ltd

bandwidth (at $-3\,\text{dB}$) of the latter, $B_{pant} = f_0/Q_1$ (see appendix), is at least capable of passing the frequencies (sidebands) contained in the signal modulating the carrier (see *Figure 3.1*).

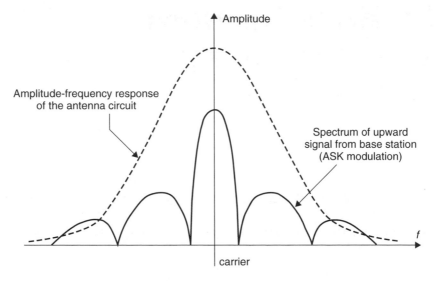

Figure 3.1 Spectrum transmitted by the base station

In the worst case (corresponding to the maximum value, $Q_{1\,\text{max}}$, of the antenna circuit of the base station), the fundamental frequency (bit rate) of the binary signals (assumed to be square, with a mark-space ratio of 50%) of this stream of digital data must therefore be at least equal to half of the bandwidth B_{pant} of the tuned circuit, in other words

$$2\,bit_{\text{rate}} = B_{pant\,\text{min}} = \frac{f_0}{Q_{1\,\text{max}}}$$

and therefore

$$Q_{1\,\text{max}} = f_0 \frac{1}{2\,bit_{\text{rate}}}$$

To make matters clearer, let us look at some examples based on commercially available components.

Example at 125 kHz
The "HITAG" component group

— Carrier frequency: $f_0 = 125\,\text{kHz}$.

In the case of applications operating with HITAG 1 and 2 circuits, for a mean digit rate of 5.2 kb/s, the duration of the pause Tp due to the 100% ASK modulation of the carrier frequency can be freely specified between 32 and 80 μs, that is, a value of $Q_{1\,\text{max}} = f_0 Tp$ between 4 and 10.

Examples at 13.56 MHz
The "MIFARE" family

— Carrier frequency: $f_0 = 13.56\,\text{MHz}$;

— digit rate of the data transmitted by the base station: $bit_{\text{rate}} = 106\,\text{kb/s}$.

In this case, we find $Q_{1\ \text{max}} = 64$.

Unfortunately, as shown in a previous book, the signal modulating the carrier has a mark-space ratio which is far from 50%, in order to provide maximum energy transfer. In the case of circuits of the MIFARE group (or the ISO 14 443 Part A), the bit coding is of the modified Miller type, in which, for a bit time of 9.44 μs, a carrier "pause" with a duration of $Tp = 3\,\mu\text{s}$ is provided. If this pause lasts for $Tp = 3\,\mu\text{s}$, it is equivalent to the transmission of a frequency whose equivalent period would be $2Tp = 6\,\mu\text{s}$, and therefore bit_{rate} is as follows:

$$\frac{1}{2Tp} = 166\,\text{kb/s},$$

and therefore

$$Q_{1\ \text{max}} = f_0 \frac{1}{2\dfrac{1}{2Tp}} \Rightarrow Q_{1\ \text{max}} = f_0 Tp,$$

that is,

$$Q_{1\ \text{max}} = 40.68 \text{ instead of 64 as stated before.}$$

This maximum theoretical value is unacceptable, since all systems have tolerances and drift (L, C, frequencies, etc.). Measures must therefore be taken to make the systems secure in all the relevant cases, assuming that in industrial conditions we can use $Q_{1\ \text{max usable}}$ of the order of 35.

The "I·CODE" component group

— Carrier frequency: $f_0 = 13.56\,\text{MHz}$.

In the case of applications operating with the "long-distance" transponders of the I·CODE group (or of the ISO 15 693), the shortest time present in the uplink communication protocol is a "pause" having the duration $Tp = 9.44\,\mu\text{s}$.

In this case, we find $Q_{1\ \text{max}} = 128$ reduced to $Q_{1\ \text{max usable}} = 100$ for the same reasons as before.

NOTE

Generally, these values of $Q_{1\ \text{max usable}}$ are not too difficult to obtain. This is because the conventional technologies used to produce antennae provide intrinsic $Q_{1\ \text{max}}$ (in the idle state) that are much higher than those required by the applications. An easy way of reducing these values to optimal values will be described in a few paragraphs.

So much for the maximum value of the quality factor required in the base station antenna. Let us now consider the parameters affecting the cut-off of the carrier during the "pauses" in 100% ASK and the rapid variations of modulation in 10% ASK. To do this, we need to examine the pulsed response of the antenna circuit.

3.2.2 Pulsed response of the base station antenna

The pulsed response (or transient response) to a voltage step E of a series RLC circuit, called a "damped resonant circuit" – see *Figure 3.2* (e.g. that of the base station antenna) – is well known and described in all basic electronics textbooks, which you should consult for further information (also see the end of the previous chapter).

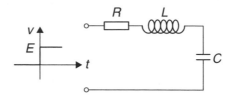

Figure 3.2   Voltage steps applied to the base station antenna

The equation for the current flowing in the circuit during the transition, in the form of the Carson–Laplace transform, is as follows:

$$I = \frac{E}{L} \cdot \frac{p}{\omega_0{}^2 + 2\Delta p + p^2} = \frac{E}{L} \cdot \frac{p}{\omega_0{}^2 \left(1 + \dfrac{2\Delta}{\omega_0{}^2} p\right) + \dfrac{1}{\omega_0{}^2} p^2}$$

To summarize, letting $2\Delta = R/L$, and assuming that

$R < \sqrt{4L/C}$ – the condition for the presence of damped oscillations –

$$\omega_0 = \frac{1}{\sqrt{LC}} \Rightarrow LC\omega_0{}^2 = 1$$

$$= 2\pi f_0$$

$$T_0 = \frac{1}{f_0}$$

When a pulsed transition is applied to what is known as a damped oscillating circuit (e.g. in ASK modulation of 100% or m%), the time response of the RLC circuit to this signal appears as an exponential voltage variation (a decrease or increase according to the polarity of the variation) of the amplitude of the oscillations having the period T. The mathematical expression of these variations, to several second-order approximations, takes the form

$$v(t) = V_{\max} e^{-\Delta t} \cos \omega t$$

where $e^{-\Delta t}$ is called the "logarithmic decrement". By way of example, *Figure 3.3a* shows the waveform when the excitation ceases (in the case of 100% ASK). As this equation indicates, the variation of voltage across the terminals of the inductance of the oscillating circuit is a cosine wave having an exponential envelope whose time

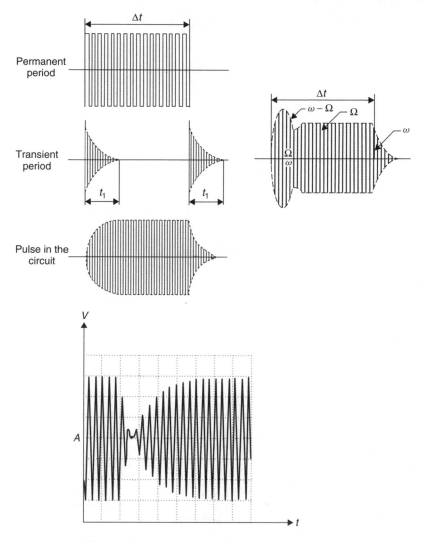

Figure 3.3a   Pulsed response of the tuned circuit

constant θ is equal to

$$\theta = \frac{1}{\Delta} = \frac{2L}{R}$$

The above equation can then be written as

$$v(t) = V_{\max}e^{-\frac{t}{\theta}}\cos\omega t.$$

Given that, for the base station antenna, $Q_1 = (L_1 s\omega)/(R_1 s)$, we can also write

$$\theta = \frac{Q_1}{\pi f_0}$$

$$= \frac{Q_1}{\pi} T_0,$$

that is,

$$\frac{\theta}{T_0} = \frac{Q_1}{\pi}$$

(see also the end of the previous chapter),

$$\Delta = \frac{1}{\theta} = \frac{\pi}{Q_1 T_0},$$

and assuming that $n = Q_1/\pi$, we obtain

$$\theta = nT_0$$

and

$$Q_1 = n\pi \quad (Figure\ 3.3b).$$

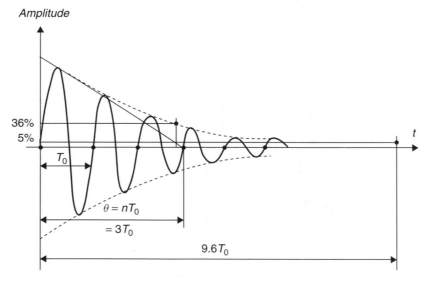

Figure 3.3b   Detail of the pulsed response

The above general equation can be used to calculate the voltages at the instants $t_1 = \theta$ and $t_3 = 3\theta$. We therefore find

—   for $t_1 = \theta$,

$$v(\theta) = V_{max}e^{-1},$$

that is,

$$v(\theta) = \frac{V_{max}}{2.718}$$

or alternatively

$$v(\theta) = 36\% \text{ of } V_{max}$$

— for $t_3 = 3\theta$

$$v(3\theta) = V_{\text{max}}e^{-3},$$

that is,

$$v(3\theta) = \frac{V_{\text{max}}}{20}$$

or alternatively

$$v(3\theta) = 5\% \text{ of } V_{\text{max}}.$$

This means, in this last case, that after a time t equal to 3θ, the amplitude of the signal is now only 5% of its initial maximum amplitude.

If the base station operates with 100% ASK amplitude modulation, this time $t3 = T_{\text{cut-off at } 3\theta} = 3(1/a) = 3(2L/R)$ is the time required for the carrier to be physically cut off (or almost, to the nearest 5%).

At the operating frequency of the circuit (f_0, $T_0 = 1/f_0$, $\omega_0 = 2\pi f_0$), we can also put this in numerical form $T_{\text{cut-off } x\%}$ as a function of the quality factor of the circuit at this frequency, $Q_1 = (L\omega_0)/R$, giving us

$$T_{\text{cut-off } 36\%} = 1\theta = \frac{Q_1}{\pi f_0} = \frac{Q_1}{\pi} T_0$$

$$T_{\text{cut-off } 5\%} = 3\theta = 3\frac{Q_1}{\pi f_0} = 3\frac{Q_1}{\pi} T_0.$$

Let us take two examples, one operating at 125 kHz and the other at 13.56 MHz, each having a quality factor Q_1 of 10, that is, for $Q_1/\pi = 3.2$ and for $3(Q_1/\pi) = 9.6$.

Carrier frequency: f_0	125 kHz	13.56 MHz	Notes
$T_0 = \dfrac{1}{f_0}$	8 µs	73.75 ns	
where $Q_1 = 10$			
if $n = \dfrac{Q_1}{\pi} = 3.2$			
$T_{\text{cut-off } 36\%}$	25.4 µs	234.8 ns	that is, approximately 3.2 carrier cycles
if $n = 3\dfrac{Q_1}{\pi} = 9.6$			
$T_{\text{cut-off } 5\%}$	76 µs	704.6 ns	that is, approximately 9.6 carrier cycles

It should be noted that the value of $T_{\text{cut-off } 5\%}$ at 13.56 MHz in this example conforms to ISO 14 443, Part 2A.

NOTES

(a)   If, for example, $Q_1 = 10$,

— The bandwidth at -3 dB for a system operating at 125 kHz is 12.5 kHz (i.e. ± 6.25 kHz), giving rates of the order of 4 to 5 kilobaud, without a subcarrier.

— The bandwidth at $-3\,\text{dB}$ for a system operating at $13.56\,\text{MHz}$ according to ISO 14 443 is $1.35\,\text{MHz}$ (i.e. $\pm675\,\text{kHz}$), which only very slightly attenuates the subcarrier at $847.5\,\text{kHz}$ of the downlink coded with the "Manchester subcarrier code" or in "BPSK", located slightly out of the band (see Figure 3.4).

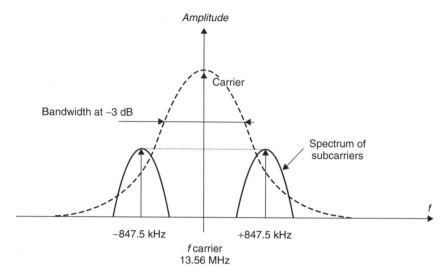

Figure 3.4   Positions of sidebands produced by subcarrier frequency

(b)   To simplify the calculation, the value at 1θ (36%) is often used for the determination of the value of Q_1.

(c)   In some cases, the quality factor is reduced electronically, using a damping transistor connected in the antenna circuit and activated only during the pause time.

(d)   If the application had required that the 5% cut-off time should be present only after three carrier cycles (i.e. $3T_0$), it would have been necessary that

$$3T_0 = T_{\text{cut-off 5\%}}$$
$$= 3\frac{Q_1}{\pi}T_0,$$

that is,

$$\frac{Q_1}{\pi} = 1$$

or alternatively

$$Q_1 = \pi = 3.14$$

which, for many other reasons (such as the need to optimize energy transfer), is considered too low.

(e)   It should be noted that all of these remarks concerning the values of Q_1 are applicable to both the 100% and the 10% ASK modulations, and relate to the best way of obtaining the correct forms of the signals for pauses and transitions from "1s" to "0s" in the worst case, for example, in continuous sequences of 10101010...

At this point in the progress of a project, the value of Q_1 of the base station antenna for the operation of the system has been defined.

This concludes the review of the information that you will need to understand the following sections relating to the driver stage.

3.2.3   Structure of the tuned LC load: series or parallel?

As we have seen, the antenna load always consists of a tuned LC circuit.

One of the major philosophical questions in the field of contactless applications is whether it is better to produce a series tuned circuit or a parallel tuned circuit, the latter also being frequently referred to as an anti-resonant circuit.

We will not burden the reader with all the well-known information about tuned circuits, but will summarize it here (see *Figures 3.5*).

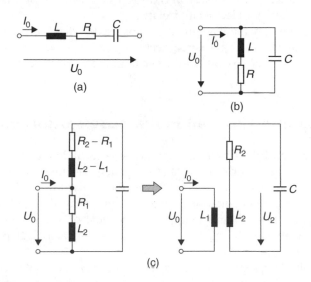

Figure 3.5   Possible configurations of electrical loads of antennae

"Series" configuration (Figure 3.5a)   In resonance, in an LC series circuit, the current – I – is "large" and is limited only by the value of the series resistance $R_1 s$ of the antenna coil, which also means that the voltage applied to the load represented by the series LC circuit is equal to the product of $R_1 s \cdot I$.

Everything depends on the current I, which is required to flow in the antenna in question, but in general, owing to the low value of $R_1 s$, this pd is not very high.

"Series/parallel" configuration (Figure 3.5b)   In resonance, in a series circuit ($L_1 s + R_1 s$), connected in parallel with a capacitance C, acting as an anti-resonant circuit, the current I supplied by the external generator is "small", and its value is limited only by the value of the parallel equivalent resistance ($R_1 p = Q_1 s^2 R_1 s$) of the antenna coil, which is high if the value of $Q_1 s$ is large. This also means that the voltage across the terminals of the parallel LC circuit is equal to $Q_1 s^2 \cdot R_1 s \cdot i$ or alternatively $L_1 s \cdot \omega \cdot I_{\text{ant}} = L_1 s \cdot \omega \cdot Q_1 s \cdot i$.

For a frequency ω with the same $I_{ant} = Q_1 s \cdot i$, everything therefore depends on the value of $L_1 s$!

The use of this kind of circuit is preferable when there has to be a high current flowing through the antenna coil ($I_{ant} = Q_1 s \cdot i$) and the inductance can be low so that the drive voltage is "relatively moderate".

Conversely, in a first approximation, we can conclude that when the inductance of the antenna is high it is preferable to use a series configuration of the LC circuit of the antenna.

"Series/parallel" configuration using a tap or transformer   Many variant layouts can be derived from the two options outlined above. *Figure 3.5c* shows two of these variants that, in fact, are the same thing.

In *Figure 3.5a*, a tap has been made on the winding of the antenna coil; because of the coupling between turns, this is equivalent to producing an autotransformer. *Figure 3.5b* shows an antenna made in the form of a true transformer in which the primary and secondary are electrically isolated.

Clearly, these two very familiar circuits can be used to distribute the ratios of values between voltages and currents as desired, according to the chosen transformation ratios.

In view of this, we need to consider the maximum value of the current in the antenna driver circuit.

3.2.4   Maximum current in the base station antenna

The maximum value of the current flowing in the base station antenna depends solely on the application and performance of the (integrated) antenna driver circuit.

In order to provide the magnetic induction (the well-known μT) required by the transponder for correct operation at the range required by the application, we must generate a magnetic field, H that is directly dependent on the values of N_1 (number of turns on the antenna), I_1 (current flowing in the antenna) and r_1, the radius of the antenna.

Briefly, in a first approximation, for a given field H the values of the three elements N_1, I_1 and r_1 can be specified freely, according to the components and systems available, as listed in the preceding section.

However, these three parameters are interrelated with the values of the inductance of the antenna ($L_1 = f(N_1$ and $r_1)$), the quality factor Q_1 of the antenna ($Q_1 s = f(L_1 s$ and $R_1 s)$), $R_1 s$ being the resistance of the antenna, which itself depends on N_1, and r_1. In other words, each factor affects all the rest!

Moreover, in resonance, the circuit $R_1 s$, $L_1 s$, C of the antenna acts as a pure resistance and therefore consumes a power P_1:

$$P_1 = R_1 s I_1^2$$

provided by the power amplifier of the base station.

Given also that

$$Q_1 s = \frac{L_1 s \cdot \omega}{R_1 s}$$

we can transfer the value of R_1s from the first to the second, giving the important relation

$$I_1 = \sqrt{\frac{P_1 Q_1}{L_1 \omega}}$$

This last equation often makes it possible to approach the problem from the other direction; in other words, if the maximum available power (meaning the maximum that can be delivered by the base station's integrated power circuit) is known, it is possible to estimate the maximum current of the base station antenna, and thus to produce a base station antenna whose inductance is compatible with the number of turns N_1 desired with a radius r_1 suitable for producing the electrical field H required for the application.

You are now almost ready to solve the problem, but be careful: there is more than one kind of maximum current! The aim of the next section is to reveal some hidden factors relating to this "maximum current".

3.2.5 Selectivity of the antenna circuit

The antenna circuit of the base station, formed by the tuned serial or parallel RLC circuit, theoretically has a quality factor Q which is unavoidably associated with two important parameters, namely, the bandwidth and the pulsed response of the antenna circuit, which have already been described in detail. This also means that the tuned antenna circuit clearly forms a bandpass filter for both transmission and reception.

Transmission   Theoretically, in transmission (in the uplink), owing to the ASK modulation of the HF carrier by the baseband signal (having a form close to a square signal), the spectrum of the latter is transposed into HF (subject to the type of modulation). In reality, where the HF current in the antenna is concerned, the HF signal is greatly altered and very different from what was wanted, because of the bandwidth of the tuned circuit of the antenna, which in this case, unfortunately, does not have a very large bandwidth.

Let us look at the example of 100% ASK modulation carried out by means of a pure square voltage (signal).

In the baseband, the Fourier analysis of this signal (see *Figure 3.6a*) gives us

$$U_{\text{ant}} = U_0 \frac{4}{\pi} \left(\frac{1}{1} \sin(\omega t) + \frac{1}{3} \sin(3\omega t) + \frac{1}{5} \sin(5\omega t) + \cdots \right).$$

As the above equation shows, the amplitude of the spectral component of order N has the following value:

$$U_{\text{ant } N} = \frac{4}{\pi N} U_0$$

The last two equations indicate that the signal consists of an infinite series of sinusoidal signals whose frequencies are odd multiples of the order $N = 1, 3, 5, 7, \ldots$ of

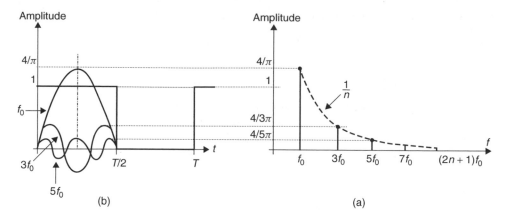

Figure 3.6   Spectral content of a square signal

the fundamental frequency of the original signal and whose fundamental frequency amplitude is greater by a factor $4/\pi$ than that of the initial square signal (all the other harmonics contribute to the reduction of this peak value, finally producing 100% of the maximum amplitude of the initial square signal. See *Figure 3.6b*).

Combining these two equations and calculating the modulus of the result, we obtain

$$|I_{\text{ant } N}| = \frac{4}{\pi N} \cdot \frac{U_0}{|Z_{\text{ant}}|}$$

If the circuit is perfectly tuned, the current of the first order harmonic (the fundamental frequency) is as follows:

$$|I_{\text{ant } 1}| = \frac{4}{\pi} \cdot \frac{U_0}{R}$$

It will therefore be necessary to take into account the correction factor $4/\pi$ to estimate the value of the maximum current which the output stage of the base station must be able to supply.

When this value has been applied to the tuned antenna circuit, owing to the high Q_1 ($Q_1 > 10$) and the associated bandwidth (Bp @ $-3\,\text{dB} = f_0/Q_1$), most of the upper harmonics of orders 3, 5 and 7 will soon be placed outside the band and therefore highly attenuated. The energy of the square signal is then principally (almost exclusively) concentrated in the first harmonic or harmonics of the radiated spectrum.

Reception   In this chapter, for the time being, we shall forget about the problem of the selectivity of the base station antenna circuit and the choices and compromises associated with the specification of its value, since there are numerous interacting factors involved here. We shall develop this theme further in the following chapters and examples, highlighting the role of each of the parameters that have a more or less direct effect on the selectivity of the antenna circuit. Please bear with the uncertainty for the time being; your patience will be rewarded in due course.

3.3   Structure of the Driver Stage of the Base Station Antenna

Regardless of whether the ASK modulation is 100% or 10%, the electronic structure of the driver stage of the base station antenna is (generally) not very complicated.
   It usually consists of

— either a conventional push–pull stage,

— or a differential push–pull stage or bridge stage, also known as an "H-bridge",

— or a stage operating in class C,

with the signal applied to the terminals of the (oscillating) output circuit.

3.3.1   Conventional push–pull stages

Figure 3.7a shows the structure of this type of output stage.

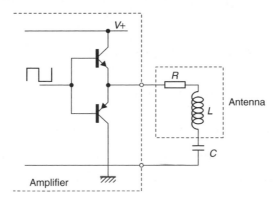

Figure 3.7a   Conventional push–pull power stage

   It is extremely simple, and has an inherent drawback, namely, that the output from this stage (the hot spot) is asymmetric with respect to earth. Therefore, although the current flows quite correctly through the antenna and produces a magnetic field, the voltage applied to the antenna is developed only at the tip of the antenna and also produces an electrical field which is propagated through the surrounding area and can potentially pollute it.
   This type of structure can therefore be recommended for systems used for short or very short-range applications, in which the energy used is low and the radiation standards (FCC, ETSI, etc.) can easily be complied with.

3.3.2   Differential or balanced push–pull stages

The characteristics of differential push–pull stages or H-bridge stages, such as those shown in *Figure 3.7b*, make them more useful for driving the antenna circuit.

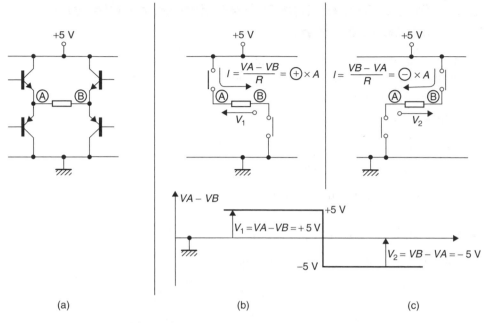

Figure 3.7b   Balanced "bridge" power stage

With these types of configuration, it is possible

—   to double the drive voltage applied to the antenna circuit,

—   to balance the antenna drive signals with respect to earth, and therefore

—   to reduce in structural terms the electromagnetic radiation emitted by the output stage.

These circuits are therefore to be recommended when the energy used is higher, in order to facilitate the achievement of greater operating ranges, without any departure from the radiation standards (FCC or ETSI).

NOTES

—   In order to balance the transmission of the signals with respect to earth, it is also possible to use HF transformers having their centre points connected to earth, or autotransformers (of the balun type, whose operation is explained below).

—   Finally, we must note in passing that the structure of the output power stage also has a non-negligible effect on the Q in real load conditions of the oscillating antenna circuit as a function of its equivalent output impedance (or resistance) (with an order of magnitude of $R_{driver} = 3.5\ \Omega$). Starting from the simple, but not excessively simple, assumption that U is sometimes equal to RI, it is often easy to estimate rapidly the order of magnitude of this equivalent resistance by measuring, at the output of the power stage, the voltage drop between the "idle" and "loaded" states, due to the current flowing in the antenna circuit.

3.4 Matching the Source Impedance (Amplifier) and Load Impedance (Base Station Antenna Circuit)

Generally, the global impedance represented by the load of the LC circuit produced in this way is not matched in anyway (in terms of power) to the output impedance (real value only) R_{out} of the power amplifier driving the antenna.

Generally, in contactless applications, the value of the load quality factor of the global antenna circuit must be average (of the order of 10 to 40), although the antenna coil has a series resistance Rsc with a very low ohmic value and therefore a high intrinsic Q. Consequently, it is almost always necessary to connect a resistance R_{ext} in series, to limit the quality factor of the system and thus simultaneously reduce the driver stage current. The direct result of this is that the reduction in current reduces the operating range.

By using special impedance matching circuits, it is possible on the one hand to limit the current in the driver circuit, and on the other hand to retain the high value of the intrinsic quality factor Q of the antenna.

In theory, the input impedance of the "antenna + impedance matching" unit thus formed can be adjusted to any value, according to the most suitable values for the planned system:

— for systems operating at 125 kHz, the value of this impedance depends mainly on the maximum value chosen for the driver stage current, as specified by the designer of the system;

— at 13.56 MHz, the same considerations apply but to an even greater extent, for reasons to do with HF applications (use of standard connecting cables, etc.); generally, the impedance is fixed at 50 Ω so that it can also be matched in power terms to the output impedance of the amplifier R_{out} and to the load represented by the whole antenna network (the capacitance bridge and the tuned antenna).

For the usual short-distance systems and applications (short range and proximity) using output stages which are either unbalanced or in H-bridge form, this impedance matching can be achieved easily, particularly by means of a network known as a "capacitance bridge".

For long-distance systems, this impedance matching is often carried out by means of a system using a transformer (of the balun or other type).

Some remarks on these different methods now follow.

3.4.1 The capacitance bridge for impedance matching: its secrets and mysteries

The circuit of what is known as a "capacitance bridge" for impedance matching is shown in *Figure 3.8*.

The equation for the complex input impedance Z_{in} of the network formed in this way, in other words the capacitance C_1 in series with $(R_1 s, L_1 s)$, itself in series and

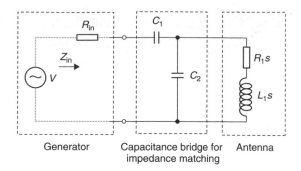

Generator   Capacitance bridge for   Antenna
impedance matching

Figure 3.8 Capacitance bridge for impedance matching

brought into parallel with C_2, is as follows:

$$Z_{in} = \frac{1}{jC_1\omega} + \frac{\dfrac{1}{jC_2\omega} \cdot (R_1s + jL_1s \cdot \omega)}{\dfrac{1}{jC_2\omega} + (R_1s + jL_1s \cdot \omega)} = a + jb$$

When the whole circuit is in resonance, the following conditions must be present simultaneously for this impedance matching to be complete:

— the value of the input impedance Z_{in} must be purely real, in other words the phase value of the impedance must be 0 degrees;

— the value of the modulus of $Z_{in} - R_{in}$ must be equal to the value R_{out} required by the designer of the application (e.g. at R_{out} of the amplifier to provide the best transferred power matching).

This means that it must be possible to write

$$Z_{in} = a + jb = R_{in},$$

that is,

— on the one hand $a = R_{in}$;

— and on the other hand $b = 0$.

The complete expansion of the equation, Z_{in} and the allowance for the last conditions for a and b, enable unique values to be determined for the capacitances C_1 and C_2 as a function of the other elements. I shall spare you the lengthy intermediate equations, but without any approximation (!) we find

$$C_1 = \frac{1}{\omega} \cdot \frac{\sqrt{R_1s}}{\sqrt{R_{in} \cdot (R_1s^2 - R_1sR_{in} + L_1s^2 \cdot \omega^2)}}$$

and

$$C_2 = \frac{R_{in} - R_1s}{L_1s \cdot \omega^2 R_{in} + \dfrac{R_1s}{C_1}}$$

In the applications with which we are concerned, since the value of ω is always high, the term $L_1 s^2 \cdot \omega^2$ is generally very large compared to $R_1 s^2 - R_1 s\, R_{in}$ and the value of C_1 is, in a first approximation, equal to

$$C_1 = \frac{1}{L_1 s \cdot \omega^2} \sqrt{\frac{R_1 s}{R_{in}}}$$

and, transferring the value of C_1 into the equation for C_2, we clearly obtain the value of C_2.

If we now consider the same circuit represented, after transformation, in its parallel equivalent form, we know that we can write

$$R_1 p = Q_1^2 R_1 s = \frac{L_1 s^2 \cdot \omega^2}{R_1 s^2} R_1 s$$

and therefore

$$R_1 p = \frac{L_1 s^2 \cdot \omega^2}{R_1 s}.$$

Carrying this into the preceding equation, we obtain

$$C_1 = \frac{1}{\omega \sqrt{R_{in} R_1 p}}$$

As regards the value of the capacitance C_2, since the value of $R_1 s$ is very low and the value of $L_1 s \cdot \omega^2$ is high, and also $L_1 s = L_1 p$, we can write

$$C_2 = \frac{1}{L_1 p \cdot \omega^2} - C_1 - Cp$$

Before we finish this section, you are invited to examine *Figure 3.9*, which makes clear the advantage of using such an impedance matching circuit, as compared with an ordinary circuit comprising only a supplementary series resistance R_{ext} to limit the drive current to a given maximum.

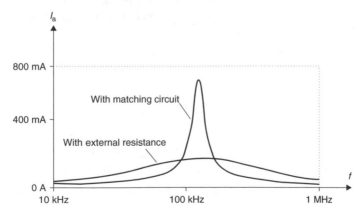

Figure 3.9 Comparison of the frequency responses of systems with and without resistance for output stage protection

As shown in this figure, the external resistance R_{ext} greatly decreases the global quality factor of the system and greatly reduces the current flowing through the antenna, whereas the proposed matching circuit has no direct effect on the quality factor of the antenna, and therefore, since it does not limit the current flowing in the antenna, allows a larger operating range without any effects on the (low) maximum drive current received from the power stage.

Special characteristics of remote antenna systems   For numerous practical reasons, it is often necessary to place the antenna at a distance ranging from a few tens of centimetres to several metres from the power amplifier that drives it.

Clearly, because of the radiation or electromagnetic pollution, this separation implies a major redesign of the physical and electronic aspects of the link between the power amplifier and the antenna. One conventional solution is to design an "all 50 Ω" system. But why "50 Ω"? The answer is simple: it is due to the fact that the link between the amplifier and antenna is generally formed by a coaxial cable whose structure is by definition unbalanced with respect to earth and whose standard characteristic impedance is 50 Ω.

Let us now examine the consequences of this choice of technology.

The consequences   First of all, while keeping the asymmetric configuration with respect to earth, the antenna winding impedance (generally low) must be matched to that of the coaxial cable (purely ohmic impedance of 50 Ω) by means of the "capacitance bridge" circuit, as described in the preceding sections. There are several possible cases here (see *Figures 3.10a* to *3.10e*):

Unbalanced output amplifier   If the amplifier output is unbalanced with respect to earth, there are two possible cases:

— either its output impedance is 50 Ω, and the problem is thus completely solved – since there is no longer any problem! (*Figure 3.10a*);

— or its output impedance is not 50 Ω, in which case a T-shaped impedance transformer must be used. Generally, in this "T" arrangement, the original L and C elements are used to provide, without any extra cost and at the same time, a harmonic filter network (*Figure 3.10b*). In terms of the circuit design, the presence of the LC harmonic filter network is completely concealed by the configuration of the impedance transformer.

Balanced output amplifier   If the output stage of the amplifier is balanced with respect to earth, a circuit of the "balun" (balanced/unbalanced) type must be used for conversion from balanced to unbalanced

— either by using a 1:1 balun transformer if the output impedance of the amplifier is also 50 Ω (*Figure 3.10c*);

— or by using a 1:1 balun transformer and a T-shaped impedance transformer system if the output impedance of the amplifier is not 50 Ω (*Figure 3.10d*);

— or by using a 1:x balun transformer if this is feasible (*Figure 3.10e*).

For further information, the principle of "active antennae" will also be discussed in Chapter 7, which deals with antennae.

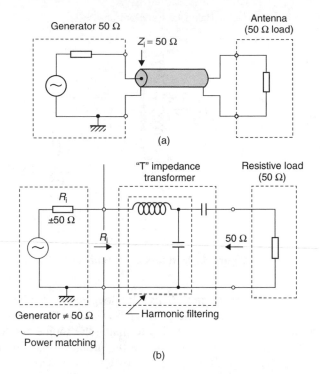

Figure 3.10a and b   Unbalanced output amplifiers

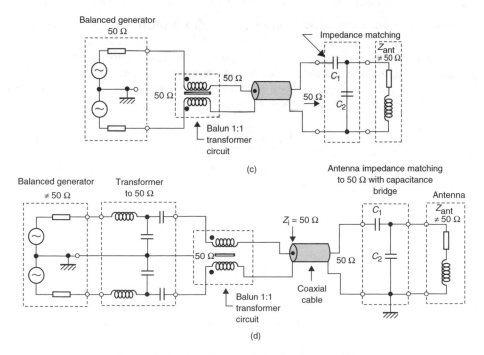

Figure 3.10c, d and e   Balanced output amplifiers

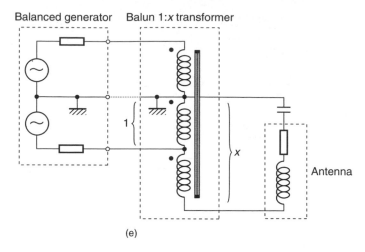

(e)

Figure 3.10c, d and e   (*continued*)

Special characteristics of long-range systems   In "long-range" applications (70 cm and above) that by definition require a higher radiated power than that found in "closed" and "proximity" systems, it is not only necessary to match the source and load impedances (e.g. 50 Ω at 13.56 MHz), but also to overcome the problems of pollution (EMC) and the associated standards (ETSI 300 330 and/or FCC 47, part 15).

To do this, we must use better-matched systems (in structural terms, this implies impedance matching circuits) with less radiation. This is what we find with balun systems.

Unbalanced to balanced conversion: the balun system   The general tendency is to use a system designed to balance the electrical radiation with respect to earth, while making the electrical signals applied to the base station antenna balanced.

This change to a balanced system from an original unbalanced system is carried out by means of a balanced autotransformer known as a "balun", having a transformation ratio $n = 2$ (step-up transformer) (see *Figure 3.11*).

$$U_{\text{secondary}} = nU_{\text{primary}}$$

Since an ideal transformer has the property of transferring the whole of the primary power to the secondary, we can write

$$\text{primary power} = \text{secondary power}$$

and therefore

$$I_{\text{secondary}} = \frac{1}{n} I_{\text{primary}}$$

so that

$$\text{secondary load seen at primary} = \frac{\text{secondary load}}{n^2}$$

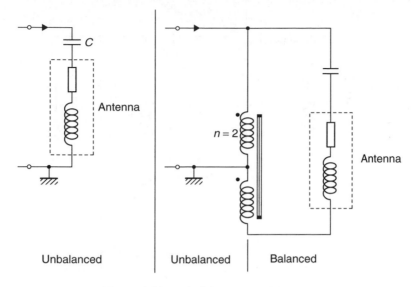

Figure 3.11 Principle of the balun circuit

For example, with a transformation ratio of $n = 2$, a load of 200 Ω will appear, seen from the input, as an impedance of 50 Ω.

The balun system and impedance matching For the reasons mentioned before, it is also necessary to match the source impedance (amplifier output) with the load impedance. Once again, a capacitance bridge circuit is used for this matching.

Thus, the global circuit including the "balun + capacitance bridge" system is as shown in *Figure 3.12*.

The physical capacitance to be added to the circuit, C_2, must also be incorporated in the new value $C'p$ of Cp and the whole load formed in this way must be found with respect to the transformer primary to estimate the load, which it represents for the amplifier ($C'p = Cp + C_2$), giving, for the values x' with respect to the primary:

$$R' = \frac{R_1 p}{n^2}$$

$$L'_1 p = \frac{L_1 p}{n^2}$$

$$C' = (Cp + C_2)n^2$$

As in the previous case, in order for the impedance matching to be complete, it is necessary that

— on the one hand, the modulus of the input impedance R_{in} of the circuit should be real and equal to R_{out};

— on the other hand, the phase of the impedance must be 0 degrees.

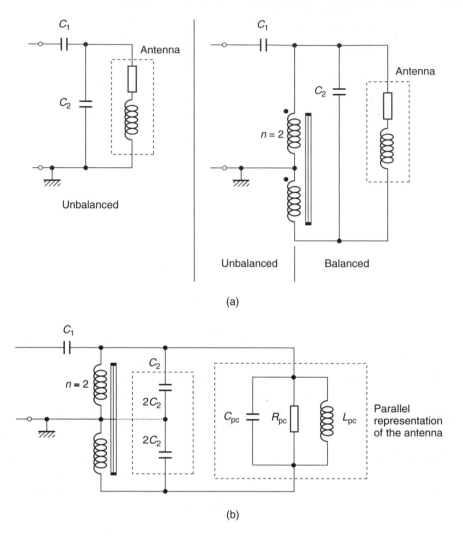

Figure 3.12   Balun circuit + capacitance bridge

To meet these conditions, the values of the two capacitances must be

$$C_1 = \frac{1}{\omega\sqrt{R_{\text{out}}R'}}$$

and, to meet the condition of resonance,

$$\frac{1}{L_1'p \cdot \omega^2} = C_1 + C' = C_1 + (Cp + C_2)n^2,$$

it is necessary that

$$n^2 C_2 = \frac{1}{L_1'p \cdot \omega^2} - C_1 - Cpn^2$$

giving us the value of C_2:

$$C_2 = \frac{1}{n^2} \cdot \left(\frac{1}{L'_1 p \cdot \omega^2} - C_1 \right) - Cp$$

Filtering and suppression of harmonic frequencies The output stage of the amplifier drives the whole "antenna unit" with a modulating signal U_{ant}, which is "square" (or "rectangular" according to the bit coding used). Now, the antenna load circuit can always be reduced, by means of an equivalent circuit, to a simple tuned RLC circuit (see *Figure 3.13a*).

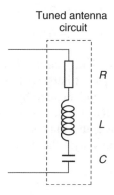

Tuned antenna circuit

R

L

C

Figure 3.13a   RLC circuit of antenna

The expression of the complex value of the current flowing in the antenna winding is as follows:

$$I_{ant} = \frac{U_{ant}}{Z_{ant}} = U_{ant} \frac{C \cdot \omega}{RC \cdot \omega + j(LC \cdot \omega^2 - 1)}$$

and, as mentioned above, after Fourier analysis, U_{ant} is written as

$$U_{ant} = U_0 \frac{4}{\pi} \cdot \left(\frac{1}{1} \sin(\omega t) + \frac{1}{3} \sin(3\omega t) + \frac{1}{5} \sin(5\omega t) + \cdots \right).$$

As the above equation shows, the amplitude of the spectral component of order N has the following value:

$$U_{ant\ N} = \frac{4}{\pi N} U_0$$

Combining the above equations and calculating the modulus of the result, we obtain

$$|I_{ant\ N}| = \frac{4}{\pi N} U_0 \frac{C \cdot \omega}{\sqrt{(RC \cdot \omega)^2 + (LC \cdot \omega^2 - 1)^2}}$$

If the circuit is perfectly tuned, the current of the first order harmonic (the fundamental frequency) is as follows:

$$|I_{ant\ 1}| = \frac{4}{\pi} \cdot \frac{U_0}{R}$$

For the other harmonics, the formula must be used. Everyone can try their hand at this, but the puzzle cannot be solved quite so easily, since at high frequencies (13.56 MHz for example) the value of R changes (increases) as a function of the order of the harmonic in question, owing to the skin effects.

Example:

at $f = 125\,\text{kHz}$; $L = 400\,\mu\text{H}$; $C = 4.7\,\text{nF}$; $R = 24\,\Omega$; $U_0 = 5\,\text{V}$
and therefore

$$|I_{\text{ant 1}}| = 129\,\text{mA}$$

$$|I_{\text{ant 3}}| = 2.5\,\text{mA}$$

$$|I_{\text{ant 5}}| = 0.85\,\text{mA}$$

If the natural filtering of the oscillating circuit due to the value of its quality factor is insufficient, second-order low-pass filters will have to be added (*Figure 3.13b*). *Figure 3.13c* (showing the circuit recommended by Philips Semiconductors for HITAG (125 kHz) and MIFARE (13.56 MHz)) is a general picture in a single figure of the balanced output of the power stage, the impedance matching and the supplementary filtering enabling all users to comply with the current standards.

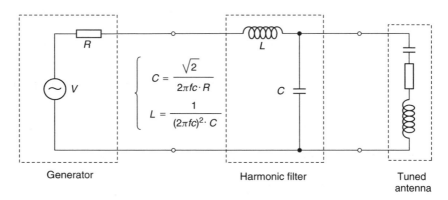

Figure 3.13b   Additional circuit for harmonic filtering of carrier

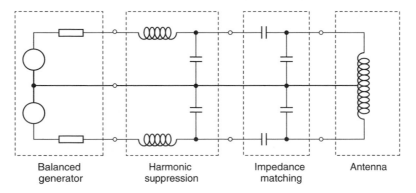

Figure 3.13c   Recommended global circuit for impedance matching and harmonic filtering

3.5   The Downlink

As far as the base station is concerned, up to this point we have only described the phenomena associated with the uplink, that is, from the base station to the transponder. We shall now examine the properties of the downlink, that is, the link from the transponder to the base station.

At the start of any design project, one of the major problems is how to estimate the voltage that the load modulation in the transponder can generate at the base station, and, following on from this, what kind of circuit should be provided or connected for processing the return signal at the base station.

We shall start by examining the theoretical problems of the level of the voltage induced at the base station by the load modulation carried out by the transponder.

3.5.1   Voltage induced in the base station antenna

After it has sent its interrogation commands, the base station switches to listening mode for the responses from the transponder.

For this purpose, the base station constantly sends a carrier and waits until the transponder signals its response by a modulation/variation of its load (*Figure 3.14*).

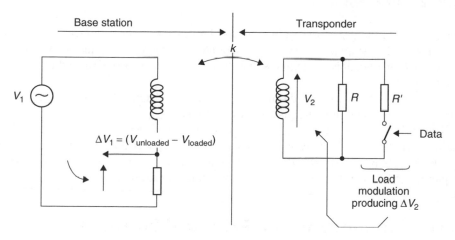

Figure 3.14   Return voltage from transponder

In terms of physics, due to the reciprocity of the mutual induction phenomenon between the primary and secondary, the voltage induced in the winding of the base station antenna is given by the formula that is a reciprocal of that which gives the voltage induced by the base station in the transponder antenna, that is,

ΔV_1 at base station antenna terminals $= \Delta V_2 M \cdot \omega I_2$ or alternatively

$$\Delta V_1 = \Delta V_2 Q_1 k \sqrt{\frac{L_1}{L_2}}$$

Clearly, this voltage decreases as the distance between the transponder and base station increases, since the coupling coefficient and the mutual inductance become lower.

For the present, to provide an idea of some of the orders of magnitude, these voltages are generally of the order of

— several hundred millivolts for proximity distances (5 to 10 cm) (k of the order of a few %);

— several hundred microvolts for long-range systems (50 to 70 cm) (k of the order of a few tenths of 1%).

Examples:

Here are some orders of magnitude:

where $L_1 = 0.5\,\mu H$; $L_2 = 5\,\mu H$; $Q_1 = Q_2 = 30$; $\Delta V_2 = 5\,V$

$$(a)\ k = 1\% \qquad \Delta V_1 = 0.5\,V$$
$$(b)\ k = 0.3\% \qquad \Delta V_1 = 150\,mV$$
$$(c)\ k = 0.01\% \qquad \Delta V_1 = 5\,mV$$

3.5.2   And then. . .

Having settled the great (meta)physical problem of the voltage induced in the winding of the base station antenna, we encounter a host of new and very important problems (amplification, detection, signal-to-noise ratio, problems related to weak collisions, etc.).

All these matters will be discussed in detail in Chapter 9 which deals with the construction of electronic circuits for base stations.

This brings us to the end of the review and supplementary information. These matters are often ignored by conventional technical publications, but I believe that the detour has been worthwhile, since we can now move on to the specification and implementation of applications – and we will find this knowledge useful in future.

3.6   Summary of the Principal Formulae of Chapters 2 and 3

Here is a summary of the principal formulae used in RFID.

Value of mutual inductance M between the base station and the transponder

$$M = \mu \frac{r^2}{2(r^2 + d^2)} N_1 N_2 s_2$$

Value of the coupling coefficient between the base station and the transponder

$$M = k\sqrt{L_1 L_2}$$

Value of the voltage across the terminals of the integrated circuit of the transponder

$$V_{ic} = \frac{1}{\sqrt{\left[\left(1 - \left(\frac{\omega}{\omega r}\right)^2\right)^2 + \left(\frac{L_2 p \cdot \omega}{R_p}\right)^2\right]}} M \cdot \omega I_1$$

Value of the minimum threshold field

$$H_{dt} = \frac{\sqrt{\left[\left(1 - \left(\frac{\omega}{\omega r}\right)^2\right)^2 + \left(\frac{1}{Qp_2}\right)^2\right]}}{\omega \mu N_2 s_2} V_{ic\,min}$$

Value of the current flowing in the base station antenna

$$I_1 = \sqrt{\frac{P_1 Q_1 s}{L_1 s \cdot \omega}}$$

Part Two
Applications and implementation

RFID and Contactless Smart Card Applications   D. Paret
© 2005 John Wiley & Sons, Ltd

4

Design and Implementation of a "Contactless" Application

So much for the preliminaries; now it is time for action, in other words, the actual implementation of an application.

Believe it or not, we must start at the beginning – meaning that we must ensure that the end user specifies the application and its magnetic environment; sometimes this is anything but simple!

4.1 Specifying an Application

Although this may seem strange at first sight, it is often difficult to draw up the specification for an application. Lengthy discussions with future users are often necessary to ensure that they describe in the correct order, and in detail, the electronic, security, mechanical, magnetic, radio frequency pollution and other constraints of the environment to which the applications will be subjected.

Your first step is therefore to take a sheet of paper and write down EVERYTHING about your application and its environment, without omitting the smallest detail. The following section will help you make sure that nothing is forgotten.

4.2 Specifying the Requirements

Throughout this book, I shall mention numerous points that are examined more closely. The following lines are a brief list of the principal parameters that must be examined as a matter of urgency before undertaking any project. These are:

Physical parameters

— mechanical formats and dimensions of the transponder (card, keyring, token, badge, etc.);

RFID and Contactless Smart Card Applications   D. Paret
© 2005 John Wiley & Sons, Ltd

— available mechanical formats for the antenna of the base station (the "reader");

— the climatic environment (operating or storage temperature, humidity, etc.);

— resistance to chemical corrosion;

— metallic environment;

— magnetic environment (be careful: some metals have no magnetic properties).

Operational parameters

— minimum and maximum operating ranges (for reading and especially for writing);

— minimum and maximum permitted times for data transactions;

— single or multiple presence of transponders in the application space;

— power/energy required for correct operation;

— "batch" or individual reading, in stacks or otherwise.

Functional parameters

— read-only, or read and write;

— capacity, structure and architecture of the transponder memory;

— "locked" data One Time Programmable (OTP), modifiable data, and so on;

— whether authentication is required between the base station and the transponder(s);

— protection of the value of the data;

— protected and encrypted messages;

— protection of the communication;

— maximum data storage period;

— where multiple transponders are present, choice of the collision management method (deterministic, probabilistic, etc.);

— whether the transponders will be moving.

Standards-related parameters

— electromagnetic interference;

— conformity with standards;

— compliance with Post Telephony & Telecommunications (PTT) and human exposure regulations.

Economic parameters

— cost;

— double source;

— conformity with standards;

— and so on.

4.3 Specifying the Near Environment of the Application

To be quite clear, the "environment" is considered to be everything that affects the materials, closely or at a distance, in relation to the magnetic or electromagnetic properties, in other words, the presence of metallic bodies, magnetic bodies, particular materials (water, ice, snow, soil, concrete, etc.) and radiation (UV, etc.) that may have a direct or indirect influence on the project.

This mainly concerns the possible paths and deformations of the magnetic field lines produced by the base station antenna, and the mode of propagation of the electromagnetic waves. We should also remember that only the base station antenna is radiating, while the transponder is not radiating at all.

4.4 How to Approach an Application

Generally, the designer of an application hardly knows where to begin, given the large number of possible starting points, and the fact that everything affects everything else! Admittedly, the problem is not simple.

Surprising though it may seem, I consider that the first point to be considered is whether or not the designer is fully aware of the radiation, pollution and human exposure standards relating to the project he is embarking on. This may help to avoid an awkward setback at the end of the project, at the time of conformity testing and certification.

By way of example and as a review of earlier work (for more details, see the author's previous book on RFID), *Figure 4.1* shows the principal standard limits used in 13.56 MHz contactless applications.

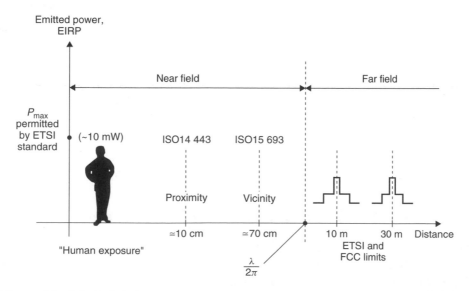

Figure 4.1 Principal limits set by standards for RFID applications. Example at 13.56 MHz

Having considered this assumption, we can now move a little further down the technical hierarchy of operations. The following section describes one procedure for working through the process at least once to obtain an initial approximation of the design of the elements. In this table, it will be necessary to introduce hypotheses, which will have to be your own rather than those developed by another designer.

4.4.1 Agenda and stages

The following lines will summarize an agenda to be followed in order to grasp an application. This procedure can be varied in numerous ways according to the specific characteristics of the applications, but the picture presented is suitable for general use.

"Analogue" part

(1) Choice of the type and performance of the transponder integrated circuit to be used for the application concerned.

(2) Determination of the electrical parameters of the transponder antenna for the best impedance matching for energy transfer.

(3) or (4) Specification of the magnetic parameters (field, induction, flux) and electrical parameters (current, voltage, power) of the output stage of the base station.

(4) or (3) Specification of the base station antenna (shape, number of turns, etc.).

"Microcontrolled" part

(5) If necessary, choice of the method (deterministic, probabilistic, etc.) to be used for collision management.

(6) Writing the specific communications protocol software for the transponder used.

(7) Writing the specific software required for the correct operation of the application.

General

(8) Tests and measurement (e.g. verification of the zero lines and the correct operation area, European Telecommunications Standards Institute (ETSI) conformity, etc.)

(9) Installation and testing on site.

(10) Finally, the opening ceremony – if everything is working correctly!

4.5 Choice of the Operating Frequency

To conclude this introduction to the implementation of applications, I shall briefly discuss a subject that is often called a "burning question" but is really only lukewarm. This is the matter of the "optimal" or "judicious" choice of the carrier frequency for the base station.

What is the best frequency for a contactless application? How many times have I been asked this question?

The answer to this question, in technical and commercial terms, depends on the nature of the application to be provided.

IMPORTANT NOTE

Before discussing this matter in the next paragraphs, may I draw the reader's attention to the fact that the following text *only relates to remotely supplied transponders*; those with built-in batteries will not be dealt with at this point. Let's avoid confusing one thing with another. Many discussions and articles in the press have been, and still are, based on the wrong premises because they make no distinction between "remotely-supplied" and "battery-assisted" systems.

While on this subject, let me make a further distinction. Confusingly, there are some transponders, particularly the SHF type (operating at super-high frequencies of 2.45 or 5.8 GHz, for example), which have built-in batteries to supply the local circuitry but whose electronic principle for communicating with the base station is called "passive" because the transponder sends out no radiation but simply acts as a "mirror" reflecting more or less of the incident radiation (back scatter).

In the following text, therefore, in order to provide real technical comparisons, I shall only discuss transponders that are really remotely supplied by the base station radiation. Clearly, in order to consider other technical performance levels, the reader may disregard these working hypotheses, but in this case he must allow for all the economic consequences and technical problems incurred by the presence of a local power supply (battery), the radio receiver and transmitter aspects, and so on, on board a transponder.

4.6 Overview of the Frequencies Used in RFID

Most systems available on the world market at present operate at one of the following frequencies or frequency ranges: below 135 kHz (125 kHz, 134.2 kHz for example), 13.56 MHz, UHF (860/960 MHz), 2.45 GHz and 5.8 GHz.

The operating and control characteristics are different for each of these frequencies, and therefore each of them is more appropriate for certain types of application or certain countries.

4.6.1 Below 135 kHz

For operating ranges up to 1 m, the 125 kHz frequency was the first to be used for high-volume industrial applications, and for short-range systems for use with car immobilizer devices. These applications are used worldwide without any special licensing. 125 kHz has become the established frequency for numerous supply chain management (SCM) applications, such as waste collection, gas cylinder and beer barrel identification at breweries, as well as for animal applications (with operation at 134.2 kHz in the last case).

4.6.2 13.56 MHz

Like the 125 kHz frequency, 13.56 MHz is already very well established in RFID applications. Hundreds of millions of products using this frequency have already been installed in many widely varying applications (access control, transport, banking, labelling, etc.). Falling within the Industrial, Scientific and Medical (ISM) band, this frequency can be used worldwide without a site licence, subject of course to the maximum authorized power limits for Equivalent Isotropic Radiated Power (EIRP).

The 13.56 MHz frequency has the following advantages:

— the base station antennae can be simple and unscreened (13.56 MHz transponders require antennae with only a few turns (3 to 8) that can be manufactured at very low cost);

— they can operate (for reading and writing) over a distance of one metre under European regulations (ETSI) and approximately 80 cm under US regulations Federal Communications Commission (FCC), providing an operating range similar to that of the 125 kHz systems.

At this frequency, depending on the ranges of the application, the data can be transmitted at a bit rate of the order of hundreds of kilobits per second (100, 400 kbit/s), in other words, much faster than in systems operating at 125 kHz, which is more than sufficient for most applications. Note that communications at 13.56 MHz are not affected by the presence of water, ice, snow or humidity. This frequency is already widely used for high-volume applications such as labels for tracing packages or baggage carried on aircraft, which may be exposed to rain, snow or condensation and water vapour, in which wet labels can be read as easily as dry ones.

The high volumes of applications such as those used in SCM, require a reliable supply of the system components, particularly the labels. In particular, the 13.56 MHz frequency has proved to be ideal for the identification of crates and products in retail SCM. Other examples can be found in postal package monitoring and rental services (books, videotapes, etc.).

The 13.56 MHz frequency is supported by all the major players in the industry, such as Philips SC, Texas Inst., ST Microelectronics, Microchip, Inside Contactless, µEM, Gem+, AXALTO, Oberthur and G and D, who offer a wide range of inlets, cards, smart labels, readers and other associated products.

Standardization   The development of a standard and its acceptance by industry is a long and complex process, usually taking three to five years. Standardization is the most effective way of ensuring the availability of compatible and/or interoperable products, thus increasing simplicity and flexibility for suppliers of services and systems integrators, while also pushing down market prices.

There has been considerable progress towards the establishment and harmonization of regulations and standards for 13.56 MHz. World regulations and statutes already support 13.56 MHz systems in Europe, the United States and Japan, specifying almost identical values (ETSI 300 330, FCC 47, part 15), and these international agreements have led to a widespread acceptance of the 13.56 MHz frequency, which is becoming ideally suited to these applications.

Several ISO standards currently cover 13.56 MHz applications. These are as follows:

ISO 14 443	*contactless smart cards (proximity cards)*
ISO 15 693	*contactless smart cards (vicinity cards)*
ISO 18 000−3	RFID *for Item Management*

Specific standards for user groups have also been created, for specific applications; one example is International Air Transport Association (IATA), which recently adopted document RP 1740C – entirely based on ISO 15 693 – which is intended to specify a solution more appropriate for baggage handling management. Several suppliers of equipment for airlines and airports now stock all the standardized products required for implementing baggage monitoring in this industry.

There are numerous applications in SCM, mainly relating to the logistical control of plastic crates of goods. For example, the labelled containers sent to the distribution centre by manufacturers are recorded on the regional distribution computer. Subsequently, within the warehouse, the position of each container is automatically updated, including the value of its contents and the date of sale, which is essential information for effective logistical control.

4.6.3 UHF

In spring 2000, a response to the question of UHF frequencies appeared to be developing. Two world market leaders, EAN International and Uniform Code Council (UCC) jointly present development plans for a programme called Global Tag (GTAG) with the aim of promoting a global standard for RFID supply chain activities. It should be noted that this proposal, essentially focused on the management of pallets and containers, gives no precise answers to the question of managing packages or boxes and articles that are already open. The promoters of the GTAG project concentrate on solutions operating in the UHF band, and urgently request the world regulatory authorities to reserve a frequency band from 862 to 928 MHz for SCM applications, since only a single band (a single channel, 8 MHz wide) can be used worldwide.

EAN/UCC points out that the use of UHF frequencies offers major advantages over the other frequencies. With these systems, reading distances of the order of 2 to 3 m can be achieved, which is important for SCM applications. Sadly and importantly, despite three attempts, this has not yet become a reality except under the current regulations in the United States (the FCC permits a power P_{eirp} from $4\,W^{eirp}$ to 915 MHz). In fact, European regulations, and particularly the current French regulations (due to the history of the use of UHF bands, the Telecommunications Regulation Authority (ART) has aligned itself with the European Recommendation ERC 70 30, which for short-range devices (SRD) only authorizes a power of $500\,mW^{eirp}$ with a maximum duty cycle ratio of 10% in the band from 869.4 to 869.65 MHz), only allow, everything else being equal, a reading distance of approximately 0.7 to 0.8 m (just the same as in 13.56 MHz systems), while in Japan 952–954 MHz band is today in discussion, no UHF band has yet been assigned for these applications. Thus, at present, the "global" solution is limited to the United States, and is far from being a world standard for the purposes of the ISO and the applications concerned.

Given the potential of UHF systems and the support of the two bodies mentioned above, all systems integrators should wait patiently until these standards are accepted and the worldwide regulatory authorities agree to dedicate a worldwide band for RFID applications in the 860–960 MHz range. This may be a long wait. Standards ISO 18 000–6 took years to reach their final version, and worldwide agreement on the use of the 860 to 960 MHz band is not completely harmonized. Moreover, it will take even longer for systems meeting the standards and regulations to become available on the market. Whatever happens, the future development of these applications should be watched closely, since their fully legitimate use on the American market will have repercussions on European products, if only because of problems due to the balance of trade.

4.6.4   2.45 GHz

In this case also, the harmonization of the regulations is a major problem with microwave transmissions in the SHF band from 2.4000 to 2.4835 GHz, generally abbreviated to "the 2.45 GHz band", since the authorized radiation limits in Europe are as yet much lower than in the United States, and are even lower in France, in accordance with the requirements of ART (500 mW in a narrow band from 2.446 to 2.454 in France, as compared to a power P_{eirp} of 4 W over the whole band, with the possibility of using Frequency Hopping Spread Spectrum (FHSS), in the United States). Clearly, this enormous discrepancy means that operating ranges will be much smaller in Europe with remotely powered systems.

Moreover, the microwave frequency of 2.45 GHz (wavelength of the order of 12.4 cm) is absorbed and converted physically into heat when it encounters molecules of human tissue or water, in the same way as in a microwave oven. Although the power used in RFID applications is far below the human injury level (and conforms to the recommendations of the WHO (World Health Organization)), radiated SHF waves are effectively screened by water, making this frequency unusable for many applications. For example, labels exposed to rain, snow or water condensation suffer a significant reduction in reading distances because of this effect.

Additionally, the appearance of local radio frequency networks (Wireless LAN) and "Wireless Connectivity" (IEEE 802.11, HomeRF, HiperLan, Bluetooth, etc.), which also operate in this frequency band over longer distances (10 m and 100 m), can frequently give rise to possible interference.

So UHF and SHF may provide the solution for the future, but what are the solutions for today?

4.7   Choosing the Right Frequency

Because of the major differences between regional regulations, as mentioned above, and certain application-specific constraints, many systems require the simultaneous use of more than one operating frequency. Although the 125 kHz and 13.56 MHz frequencies can theoretically be used for most applications, some require longer operating ranges, often achieved with active systems (having built-in batteries). Although some proprietary solutions are available, for these applications we may have to wait for some years, until the standards are completed and regulations harmonized, before we see the first solutions and systems operating on a worldwide level.

4.7.1   kHz versus MHz and GHz

Without going into lengthy theoretical discussions and/or arguments, the following text provides information in summary form about the parameters which may make one frequency more attractive than another.

<135 kHz

The use of frequencies such as 125 kHz or 135 kHz is usually only justified if

— the bit rate is not too high (several kilobits/second) or not critical;

— the EMC problems are difficult to resolve;

— the operating range is medium (15 cm, 30 cm) or long (1 m);

— many transponders are present at once in the magnetic field, in stacks, in batch, and so on.

13.56 MHz

The use of frequencies such as 13.56 MHz is mainly justified when

— the transponder antenna has to be small;

— its cost must be low;

— the environment is not too "metallic";

— the bit rate is high (hundreds of kilobits/second) AND the operating range is short (10 to 15 cm);

— the bit rate is low (approximately 2 kbit/s) or medium (approximately 20 to 30 kbit/s) AND the operating range is long (approximately 1 m). In fact, meeting the FCC and ETSI radiation standards can be tricky and makes it necessary to reduce the bit rate considerably.

UHF

The use of frequencies such as those from 860 to 960 MHz, is usually only justified if

— the bit rate or the transaction time required by the application has to be fast;

— the operating range is medium, since otherwise it is difficult to comply with local regulations, or else the bit rate has to be reduced considerably;

— the antenna of the transponder has to be small and inexpensive;

— the environment is not too "metallic".

2.45 GHz

The use of frequencies such as 2.45 kHz or 5.8 kHz, is usually only justified if

— the bit rate or the transaction time required by the application has to be fast;

— in remotely powered mode, if the operating range is medium, otherwise it is very difficult to comply with local regulations;

— the antenna of the transponder has to be small and inexpensive;

— the environment is not too "watery".

Characteristic /Frequency	< 135 kHz	13.56 MHz	860/960 MHz	2.45 GHz
Frequency	Low	High	UHF bands	Hyper
Antenna technolog (+/– impact on cost)	Air-core or ferrite-core ** coil	Printed, perforated, *** etched	Printed, perforated, ** etched	Printed antenna *** Etched
Read/write range	> 1 mm **	Europe/France: 1 m ** USA > 0.8 m	Europe/France: GSM/Telecom frequencies not available at present USA > 1 m to 10 m	France < 0.1 m (with remote power supply) * (> 1 m with derogation) *** US > 1 m
Data transfer rate	< 10 kbit/s *	< 100 kbit/s **	< 100 kbit/s **	< 200 kbit/s ***
Effect of metal(°)	disturbance * (space > 50 mm = 90% perf.)	disturbance * (space > 50 mm = 90% perf.)	attenuation ** (space > 10 mm = 100% perf.)	attenuation ** (space 5 to 7 = 100% perf.)
Effect of water	none ***	attenuation **	attenuation **	disturbance *
Effect of human body	none ***	attenuation **	attenuation **	disturbance *

(°measurements for guidance only, made by ID System)

Performance: * = Mediocre
                  ** = Good
                 *** = Excellent

Figure 4.2   Summary of principal performances/characteristics of the frequencies used in RFID (at a given date!)

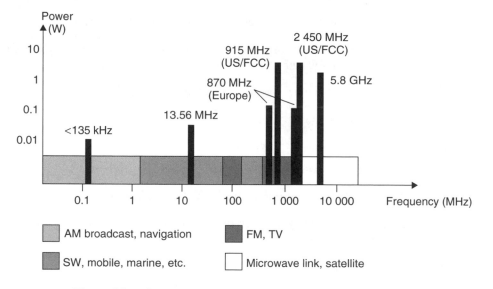

Figure 4.3a   Summary of the principal power levels used in RFID

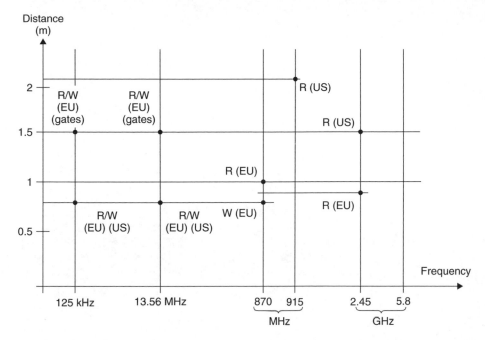

Figure 4.3b Summary of the distance/frequency options for remotely supplied transponders (at a given date!)

The table in *Figure 4.2*, based on a large amount of information from the trade, is not comparative but provides a summary of the principal performance data or specific characteristics of the use of each of these frequencies.

Figures 4.3a and *4.3b* show the authorized power ratings for different countries and the possible operating ranges.

Part Three

Examples

This third part of the book sets out, in the form of detailed examples, the knowledge gained by the author in his years of experience of contactless and RFID applications in the fields of industrial identification, immobilizers for motor vehicles, short or long distance monitoring and traceability applications, and the areas of transport, automated cash handling and access control where contactless smart cards are used.

The two chapters in this part describe in detail the technicalities of the method for setting about the theoretical specification of the principal components of the base station and transponder in contactless applications using the frequencies of 125 kHz and 13.56 MHz, which are currently most widespread in the market. No specific applications operating at UHF and at 2.54 or 5.8 GHz are described in this book, since the current European law and especially the French law on radiation standards are such that very few purely passive RFID applications can be implemented. If the development of the standards makes it possible, I will expand this area of applications in deep detail in a future book.

The approach taken in these chapters is a general one and will meet the needs of many designers, developers and businesses, regardless of the types, models and brands of integrated circuits used to build the transponders and base stations.

In spite of the general approach of this presentation, I have provided some concrete details to illustrate the points made, by choosing some specific examples of industrial components that are well known on the market, and are highly representative of various major classes of applications, namely, on the one hand access, control, transport and payment cards with operating ranges of the order of 10 cm (short range, called "proximity" by the ISO), and on the other hand widely used labels with operating ranges of the order of 70 to 80 cm (long distance, or "vicinity").

Additionally, throughout the examples, particular attention will be paid to the optimization of the specification of the components, which will lead to a considerable reduction in the cost of the units produced in this way.

While showing ways of minimizing cost and optimizing the performance of components, these chapters will also examine methods and sequences for organizing the preliminary calculations forming the basis of the actual development of a contactless system. These include the following points:

— specification of the electrical values of the various components of

    — the base station,

    — the transponder;

RFID and Contactless Smart Card Applications   D. Paret
© 2005 John Wiley & Sons, Ltd

— evaluation of the performance (in terms of current, voltage, power, etc.) required in the various components, namely,

  — the base station antenna,

  — the power driver stage of the base station,

  — specification of the transponder antenna;

— evaluation and calculation of the required values of magnetic fields and induction;

— ways of overcoming the problems of the tolerances of the various components.

A final complementary section provides detailed descriptions of the position of the calculations presented and their conformity with respect to the current standards (ISO, ETSI, FFC, etc.).

NOTE

The theory and technology of the physical construction of base station and transponder antennae (shapes, calculation of the number of turns, etc.), and the development of software for the communications protocol between the base station and the transponder, are considered to be outside the scope of these chapters, since they are dealt with in several other chapters in Part Four of this book.

Thus the following chapters relate to the preliminary work which should be carried out before a design project, and which is very rarely done by systems developers, often leading to very large discrepancies in the performance and reproducibility of contactless devices, since they are all too often developed in a trial-and-error way.

I will of course be referring frequently to the earlier chapters on theoretical calculations – which explains their position in this book.

FINAL NOTE

I have no intention of restricting the application of common sense, and therefore every user's experience will supplement these theoretical preliminary calculations with the information required to solve any given problem, allowing for specific environmental parameters of the implementation that are often very difficult to describe in theoretical terms.

5

Examples at 125 kHz

5.1 The Usual Constants and Parameters of Applications Operating at 125 kHz

Table 5.1 summarizes the principal constants and parameters normally used in contactless applications operating at 125 kHz.

5.2 Example

Let us look at the first example:

We will assume that an access control system of the voluntary action type is to be implemented; in other words, the person requiring access has to present his badge, with a transponder fitted, to the "reader" (the base station), so that, in this kind of application, the operating range remains practically constant, at approximately 5 to 10 cm.

Table 5.1

$$fc \qquad\qquad = 125\,\text{kHz}$$

$$\omega c \;\; = 2\pi fc = 785 \times 10^3\,\text{rad/s}$$

$$\omega c^2 \qquad\qquad = 61.6225 \times 10^{10}\,(\text{rad/s})^2$$

$$Tc \;\; = \frac{1}{fc} \quad\;\; = 8\,\mu\text{s}$$

$$v \qquad\qquad\;\; = 3 \times 10^8\,\text{m/s}$$

$$\lambda c \;\; = vTc \;\; = 2400\,\text{m}$$

$$\lambda c^4 \qquad\qquad = 33.1776 \times 10^{12}\,\text{m}^4$$

– Constant of radiation resistance of an antenna

$$\text{const} \quad = \frac{31\,200}{\lambda c^4}$$

$$= \frac{31\,200}{33.177 \times 10^{12}}$$

$$= 940.4 \times 10^{-12}$$

RFID and Contactless Smart Card Applications   D. Paret
© 2005 John Wiley & Sons, Ltd

5.2.1   The transponder

Let us assume that, for specific operational reasons (memory capacity, security, etc.) we have decided to use the HITAG integrated circuit produced by Philips Semiconductors, operating at 125 kHz (the reasoning followed here is of course adaptable and applicable to other transponders available on the market).

The bit rate which can be obtained with this circuit is approximately 2 kb/s. As indicated previously, this implies a value of $Q_2 s = Q_2 s_{max} \approx 32$.

Since the range for this type of application is neither very short nor very long, the voltage applied to the integrated circuit is such that the value of R_{ic} is of the order of 60 kΩ.

For an optimal energy transfer between the antenna circuit of the transponder and the load represented by the input impedance R_{ic} of the transponder's integrated circuit, the optimal inductance of the antenna circuit, as shown in Chapter 2, must be

$$L_2 s = \frac{R_{ic}}{\omega_c \cdot Q_2 s}$$

$$= \frac{60 \times 10^3}{2 \times 3.14 \times 125 \times 10^3 \times 32}$$

$$= 2.39 \, \text{mH (produced in the air or in a ferrite rod)}$$

and therefore, in order to tune the transponder globally, we must provide a global tuning capacitance C_2 such that

$$L_2 s C_2 \cdot \omega_c^2 = 1$$

that is,

$$C_2 = 675 \, \text{pF}$$

This global value C_2 includes the input capacitance of the integrated circuit (of the order of 15 to 20 pF).

Clearly, for practical reasons, the closest standard capacitance is generally chosen for the global tuning of the transponder (e.g. 680 pF, which in the present case would give a higher total capacitance of approximately 700 pF), and it is necessary to slightly adjust the value of L_2 accordingly to obtain the desired tuning.

NOTE

Integrated circuits very often have a built-in capacitance of 210 pF and therefore an external capacitance of 470 pF must be added.

At this stage of the procedure, Q_2 has been chosen and the values of L_2 and C_2 have been determined.

5.2.2   The base station

For the analogue part of the base station, we will use a system built around the HT RC 110 base station integrated circuit made by Philips Semiconductors

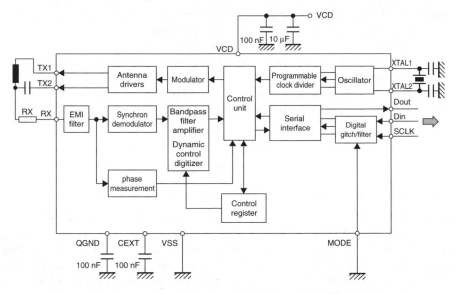

Figure 5.1 Base station built round the Philips SC-HT RC 110 circuit

(see *Figure 5.1*) whose output stage is constructed with a bridge circuit (for further details, see Chapter 3).

The principal characteristics of this integrated circuit are as follows:

$$\text{continuous supply voltage} \qquad U_0 \quad = 5\,\text{V}$$
$$\text{internal resistance of the output transistors} \quad R_{\text{driver}} \quad = 2.5\ \Omega$$

In purely theoretical terms, the maximum current could be $5\,\text{V}/2.5 = 2$ A, but in fact the manufacturer states that

$$\text{maximum permanent peak current} \qquad I_{\text{peak}} \qquad = 200\,\text{mA}$$
$$\text{maximum permitted peak current for pulsed mode} \quad I_{\text{drive max}} \quad = 400\,\text{mA}$$

In a first approximation, we will not allow for the output transistor breakdown voltage and we will assume that the output voltage of the integrated circuit is a square signal with a "peak" of 5 V (see *Figure 5.2*).

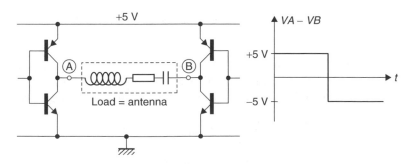

Figure 5.2 "Bridge" or "H" output stage

Given that the load circuit is a tuned circuit of the series type, formed by L_1, C_1, R_1 and the quality factor $Q_{1\,\text{max usable}}$ under load of the order of 10, ... 30, we must allow for the fact that the current flowing in the transistors of the amplifier output stage will primarily contain the first harmonic of the excitation signal that has been subjected to the familiar multiplication factor of $4/\pi$. This means that the system is considered to act in the same way as if the tuned circuit L_1, C_1 were driven by a pure sinusoidal carrier frequency for which the peak amplitude of the sinusoidal (not square!) voltage would be

$$U_{\text{drive peak}} = 5 \times \frac{4}{\pi} \text{ peak}$$

$$\approx 6.4 \text{ V peak}$$

that is, the equivalent of a sinusoidal voltage of 12.8 V peak to peak.

Output stage without protection against short circuits of the antenna winding   In the case of an output stage without protection against short circuits of the antenna winding (*Figure 5.3a*), the only outstanding problem is how to avoid exceeding, during operation, the constant maximum current value I_{drive} which the designer has decided to use for his application. In this case, we can write (for the first harmonic of the signal, hence the presence of the factor $4/\pi$) that

$$\frac{4}{\pi} \cdot \frac{U_0}{I_{\text{drive}}} = R_{\text{in}} + R_{\text{driver}}$$

an equation that completely defines the value of R_{in}.

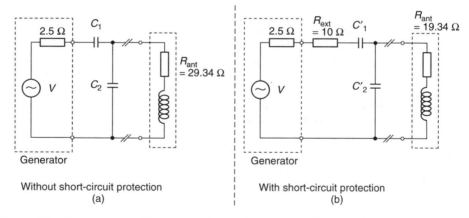

Without short-circuit protection
(a)

With short-circuit protection
(b)

Figure 5.3   Output stage without protection against antenna short circuits (a) and with protection (b)

Having assumed that the integrated circuit can constantly deliver a peak current I_{peak} of 200 mA (and therefore that the value $I_{\text{max ant}}$ is arbitrarily known), it is then possible to calculate the minimum usable value of load resistance at resonance, that is,

$$\frac{U_{\text{drive peak}}}{I_{\text{drive peak}}} = R_{\text{total minimum load}}$$

$$\frac{4}{\pi} \cdot \frac{5}{0.200} = 31.84 \ \Omega$$

This value represents the total equivalent minimum resistance (R_{total}) of the elements connected in series with the antenna, in other words, the series resistance of the antenna R_{ant} plus the ($R_{copper} + r_{loss\ in\ Fe}$ + additional resistance) and finally the amplifier output resistance(s) R_{driver}.

Given that $R_{driver} = 2.5\ \Omega$, this means that the value of R_{in} is as follows:

$$R_{in} = 31.84 - 2.5$$

$$= 29.34\ \Omega$$

If, for other reasons (to maximize the energy supply, etc.), the duration of the pause of the 100% ASK modulated signal from the base station is set at $3T_0$, this means that the circuit L_1, C_1 has a quality factor

$$Q_1 = \frac{L_1 \cdot \omega}{R_{tot}} = \frac{\theta \pi}{T_0} = 3\pi \approx 10$$

(by the way, you should note that the bandwidth at $-3\,dB$ will be, by definition, f_0/Q_1, that is, $(125\,kHz)/10 = 12.5\,kHz = \pm 6.75\,kHz$ and that the pulsed response will be 1θ).

The peak voltage across the terminals of L_1 will then be

$$UL_{1\ peak} = Q_1 U_{drive\ peak}$$

$$= 10 \times 6.4$$

$$= 64\,V\ peak,\ that\ is,\ 128\,V\ peak\ to\ peak,\ or\ 45.26\,V_{rms}$$

Note also that the same voltage will be present across the terminals of the tuning capacitance which must therefore withstand this voltage, as well as having a precise value.

Given the specification of a 200 mA peak current, the reactance $X_1 = L_1 \cdot \omega$ of the inductance L_1 will be as follows:

$$X_1 = \frac{UL_{1\ peak}}{I_{1\ peak}}$$

$$= \frac{64}{0.2}$$

$$= 320\ \Omega = Q_1 R_{total}$$

Thus we can easily find the value of L_1:

$$L_1 = \frac{X_1}{\omega c}$$

$$= \frac{320}{2\pi \times 125 \times 10^3}$$

$$= 407\,\mu H$$

and finally, according to the standard equation $L_1 C_1 \cdot \omega c^2 = 1$, the circuit L_1, C_1 is tuned by means of the capacitance whose value is $C_1 \approx 4$ nF.

From the values $R_{out} = R_{driver}$ and R_{in}, we can calculate the values of the capacitances forming the impedance matching bridge.

NOTE

To ensure that the global antenna circuit has a quality factor of 10, the total series resistance of the circuit must be as follows:

$$Q_1 = 10 = \frac{L_1 \cdot \omega c}{R_{tot}} = \frac{X_1}{R_{tot}},$$

that is,

$$R_{tot} = \frac{X_1}{Q_1}$$
$$= \frac{320}{10}$$
$$= 32 \ \Omega$$

As it happens, the sum of resistances present in series is exactly 31.84 Ω, including 2.5 Ω of driver resistance. To obtain the required value of Q_1, the resistance of the antenna winding must therefore be 29.3 Ω, which can easily be achieved by using the appropriate wire diameter for constructing the antenna winding for the base station.

In this case, however, if the physical connections of the antenna winding are accidentally short-circuited, there will be no limit on the maximum output current of the integrated circuit – causing an abrupt termination of its life, unless …

Being very observant, you have no doubt noticed the "optional" presence of an additional resistance, and asked yourself, "What's that for?" Such a resistance (generally ranging from a few ohms to about 10 ohms, as shown in the next paragraph) is often connected at the antenna driver output stage. This resistance makes it possible to provide better current driving of the antenna, and also to adjust the value of Q_1 in load conditions and, finally, to protect the output transistors against any short circuits of the antenna connecting wires, as will now be demonstrated.

Output stage with protection against short circuits of the antenna winding   In this case, if a short circuit occurs in the physical connections of the antenna coil winding, then in order to limit the maximum current to $I_{drive\,max}$ (400 mA), it is necessary to connect in series (*Figure 5.3b*), as closely as possible to the power stage of the integrated circuit, a resistance of $R_{s\,min}$ such that

$$R_{s\,min} = \frac{U_0}{I_{drive\,max}} - R_{driver} = \frac{5}{0.4} - 2.5 = 10 \ \Omega$$

and this will ensure that the maximum current supplied by the amplifier never exceeds the crucial value of 400 mA. However, in normal operation, if we wish to retain the value of the maximum current I_{drive} shown above (200 mA), we can also write (for the first harmonic of the signal, hence the presence of the factor $4/\pi$) that

$$\frac{4}{\pi} \cdot \frac{U_0}{0.200} = R_{in} + R_{driver} + R_{s\,min} = 31.84 = R_{in} + 2.5 + 10$$

an equation that completely defines the value of $R_{in} = 19.34\,\Omega$, giving the new values of C_1' and C_2',

$$C_1' = 23.3\,\text{nF}$$

$$C_2' = 62.4\,\text{nF}$$

NOTES

— The external resistance of $R_{s\,min} = 10\,\Omega$ is provided purely in order to protect the integrated circuit against any short circuits of the antenna winding terminals.

— This circuit has an inherent drawback, namely, that the quality factor of the circuit can sometimes be modified (reduced) because of the presence of the resistance $R_{s\,min}$, which can sometimes cause a reduction in the operating range. Nothing comes free of charge!

As regards the mechanical implementation of the base station antenna (diameter, radius, side measurement, thickness, etc.), this typically depends on the shape that is desired or is required by the application (circular, square, etc.). If nothing is defined at this stage, it is impossible to determine the number of turns N_1 that the antenna must have to meet the inductance value, which is a considerable nuisance since the magnetic field H is dependent on the knowledge of N_1 and r_1.

However, you should realize that each antenna coil technology (in terms of the circular/square shape, dimension, diameter, thickness, arrangement and diameter of winding wires, etc.) has a corresponding value of Lo representing the inductance of a single turn of the base station antenna (for guidance, these particular points are discussed in detail in Chapter 8).

Given that the inductance of a conventional coil is approximately proportional to the square of the number of its turns, we can write $L_1 = L_1 o N_1^2$. From the above equation we can derive the following relation:

$$N_1^2 = \frac{U_{\text{drive peak}}}{I_{\text{peak}}} \cdot \frac{Q_1}{\omega c \cdot L_1 o}$$

and therefore

$$N_1 = \sqrt{\frac{U_{\text{drive peak}}}{I_{\text{peak}}} \cdot \frac{Q_1}{\omega c \cdot L_1 o}}$$

Clearly, if the value L_1 of the inductance of the antenna is specified, the calculations can be carried out in reverse to determine the maximum current to be supplied by the power stage of the base station, but note that, because of the last equation, even with specified values of Q_1 and $L_1 o$, there is still a certain amount of freedom in the choice of distribution of the pair U and I_1, depending on the desired implementations.

At this stage of the design process, the maximum antenna current has been chosen and the electrical values of L_1 and C_1 have been determined, together with the number of turns N_1.

Electrical power, field and magnetic induction required

Power To conclude these initial calculations, we can add that the power in watts delivered by the output stage is

$$P_{1\text{ rms}} = U_{\text{rms}} I_{\text{rms}} = R I_{\text{rms}}^2$$

and therefore, for a maximum I_{peak} of 200 mA effective,

$$I_{\text{rms}} = \frac{I_{\text{peak}}}{\sqrt{2}} = 141 \text{ mA}$$

we obtain

$$P_1 = (31.8 + 3.5) \times 0.141^2$$

$$\approx 700 \text{ mW}$$

Magnetic field If the antenna coil is circular, with a radius r_1, the value of the magnetic field in its centre is given by the formula

$$H_0 = \frac{N_1 I_1}{2r_1}$$

Replacing N_1 with its value found in the previous section, we find that the magnetic field created by the base station in its centre is as follows:

$$H_0 = \frac{N_1 I_1}{2r_1} = \sqrt{\frac{U_{\text{drive peak}}}{I_{\text{peak}}} \cdot \frac{Q_1}{\omega_c \cdot L_{10}}} \cdot \frac{I_1}{2r_1}$$

and therefore the magnetic induction is

$$B_0 = \mu H_0$$

5.2.3 Example

Finally, let us look at a specific example of implementation.

A transponder sold as a ready-made product requires a peak induction of approximately $50 \, \mu\text{T}$ to operate correctly. The proposed application is a proximity system (e.g. an access control badge or an immobilizer for a motor vehicle) for which the nominal operating range is 4 cm, and, assuming that the base station antenna is circular, flat and air-cored, with a radius of 2 cm and a quality factor Q_1 equal to 10 and an inductance of $407 \, \mu\text{H}$. To obtain this value, the antenna must have the following approximate number of turns (see Chapter 3):

$$L_1 = \frac{\mu \pi r N_1^2}{2}$$

that is,

$$407 \times 10^{-6} = \frac{4 \times 3.14 \times 10^{-7} \times 3.14 \times 0.02 \times N_1^2}{2}$$

and therefore

$$N_1 = 102 \text{ turns}$$

that is, inductance per turn (L_{1^0})

$$L_{1^0} = \frac{L_1}{N_1^2}$$

$$= \frac{407 \times 10^{-6}}{102^2}$$

$$= 39.1 \text{ nH per turn}$$

First of all, we must calculate the value of the magnetic field $H_{4 \text{ cm}} = H_{2r}$:

$$H_{4 \text{ cm}} = \frac{B_{4 \text{ cm}}}{\mu}$$

$$= \frac{50 \times 10^{-6}}{4 \times 3.14 \times 10^{-7}}$$

$$= 39.8 \text{ A/m peak}$$

Given that, in this example, the operating range d is equal to $2r$, or alternatively $H_{4 \text{ cm}} = H_{2r}$, we can then easily calculate H_0:

$$H_0 = H_{2r} \times 11.8$$

$$= 470 \text{ A/m peak}$$

Also, given that the antenna coil is circular and flat, we can write $H_0 = (N_1 I_1)/2r_1$, that is,

$$I_1 = H_0 \frac{2r_1}{102}$$

$$= H_0 \frac{470 \times 2 \times 0.02}{102}$$

$$= 0.183 \text{ A peak}$$

$$= 0.132 \text{ A rms}$$

You should note in passing that the peak value of the current in this example is compatible with the maximum value supported by the integrated circuit proposed at the beginning of the chapter.

The voltage across the terminals of the antenna will be

$$U_1 = L_1 \cdot \omega I_1$$

$$= 407 \times 10^{-6} \times (2 \times 3.14 \times 125 \times 10^3) \times 0.132$$

$$= 42.17 \text{ V rms}$$

Given that, additionally, I_1 is equal to $\sqrt{(P_1 Q_1)/(L_1 \cdot \omega)}$, the power delivered by the integrated circuit will be as follows:

$$
\begin{aligned}
P &= \frac{L_1 \cdot \omega I_1{}^2}{Q_1} \\
&= \frac{407 \times 10^{-6} \times 2 \times 3.14 \times 125 \times 10^3 \times 0.132^2}{10} \\
&= 557 \, \text{mW}
\end{aligned}
$$

Summary of the 125 kHz application

Transponder:

$L_2 s = 2.39 \, \text{mH}$

$C_2 \ = 675 \, \text{pF}$

Base station:

$L_1 s = 407 \, \mu\text{H} \rightarrow N_1 = 102$

$R_1 s = 19 \, \Omega$

$R_{\text{ext}} = 10 \, \Omega$

$Q_1 \ = 10$

$V_1 \ = 45.26 \, \text{V rms}$

$C_1{}' = 23.3 \, \text{nF}$

$C_2{}' = 62.4 \, \text{nF}$

$H_0 \ \approx 470 \, \text{A/m}$

$I_1 \ \approx 140 \, \text{mA}$

$P_1 \ \approx 600 \, \text{mW}$

6

Examples at 13.56 MHz

In this chapter, I shall present several examples of the numerous classes of contactless applications operating at 13.56 MHz.

These include

(a) very short-range (2 to 3 cm) applications, of the ISO 10 536 "closed distance" type;

(b) proximity applications (10 to 15 cm) of the ISO 14 443 "proximity" type;

(c) vicinity applications (50 to 70 cm) of the ISO 15 693 "vicinity" type, very suitable for item management;

(d) applications of the "item management" type according to ISO 18 000 part 3.

To make the chosen examples more realistic and specific, they must be quantified, using precise and specific technical characteristics.

Clearly, whatever prior attitudes we may have, the choice of components to illustrate the examples in this book is a difficult matter. I have decided to illustrate examples (a) and (b) with the integrated circuits of the MIFARE family, produced by Philips Semiconductors, since these conform strictly to ISO 14 443, and, what is more, this technology has a long history of use for these applications around the world. Examples (c) and (d) make use of SLI (Smart Label ISO) components of the I·CODE, since these conform to ISO 15 693 and ISO 18000-3.

Many other commercially available components could have been used to illustrate these examples[1]; what is essential is that potential users will find that this chapter gives all the necessary information for the operation of their specific products.

6.1 The Usual Constants and Parameters of Applications Operating at 13.56 kHz

Table 6.1 summarizes the principal constants and parameters normally used in contactless applications operating at 13.56 kHz.

---

[1] Here is a list of the main ones, in alphabetical order and without any hint of preference: Atmel, Electronique Marin, Infineon, Inside Contactless, Micro Chip, Motorola, ST Microelectronics, Texas Instruments, etc.

RFID and Contactless Smart Card Applications   D. Paret
© 2005 John Wiley & Sons, Ltd

Table 6.1

$$fc = 13.56\,\text{MHz}$$

$$\omega c = 2\pi fc = 85.157 \times 10^6\,\text{rd/s}$$

$$\omega c^2 = 7251.55 \times 10^{12}\,(\text{rd/s})^2$$

$$Tc = \frac{1}{fc} = 73.75\,\text{ns}$$

$$v = 3 \times 10^8\,\text{m/s}$$

$$\lambda c = vTc = 22.12\,\text{m}$$

$$\lambda c^4 = 2.394 \times 10^5\,\text{m}^4$$

– Constant of radiation resistance of an antenna

$$\text{const} = \frac{31\,200}{\lambda c^4}$$
$$= \frac{31\,200}{2.394 \times 10^5}$$
$$= 0.13032$$

6.2   ISO 14443 "Proximity" Applications (Approximately 10 cm)

These "proximity" applications primarily consist of "voluntary action" applications. This description means that the user makes a voluntary decision to present his transponder (contactless smart cards, access badges, etc.) to the base station.

This is mainly due to the fact that, in these applications, sensitive data (special access rights, money, etc.) are exchanged in the transactions, that the user wishes to be as discreet as possible overall, or in the last analysis, that he appears to enjoy a higher level of security because he is in regulation of the transaction and nothing can happen without his knowledge in the immediate environment.

These applications are normally made secure and protected by encoded and encrypted communications and support significant arithmetical and logical calculations.

6.2.1   The MIFARE family

The MIFARE hardware platform is well known in the market. This class includes a range of equipment from the very simple MIFARE Ultra Light circuit to the very conventional MIFARE 1, both produced in the form of wired logic based on microcontrolled architecture, to the more complex circuits such as MIFARE Pro and MIFARE Pro X, Smart MX, DESFire, all of the pure microcontroller type (the core of the 80C51 class) with a dual interface (contact and contactless). To summarize their performance in a few words:

— MIFARE Standard:
   MIFARE Standard is a special-purpose integrated circuit with 1 kB of E2PROM, developed and optimized particularly for contactless applications in public transport ticketing, which has subsequently become a de facto international standard,

having shown that its functionality, reliability and security concept are generally satisfactory for this type of application.

Subsequently, this circuit has been accepted as the preferred option for security protected access control, "city cards" for access to services and multi-application concepts using its facility for combining 16 different protected applications on one card (e.g. multi-fare multi-mode transport cards, electronic wallets, access controls, loyalty programmes, etc.).

— MIFARE Ultra Light:
MIFARE Ultra Light is a simplified version (384-bit E2PROM, single application) of MIFARE Standard.

— MIFARE Pro:
MIFARE Pro was the first member of a class of microcontrollers (improved 80C51 core) with a dual interface – contact and contactless – and a CPU optimized for efficient use of the available energy in "proximity" (10 cm) contactless mode, according to ISO 14 443 – A. This circuit meets the speed requirements for transport applications in the MIFARE Standard range, and also meets the high-security requirements of banks, which are becoming increasingly interested in contactless systems in view of the high volumes of transactions required by mass transport. Banking institutions require (initially, at least) transactions using a contact interface, with the implementation of algorithms tested by the banks themselves (DES type), both in contact and contactless mode, in an "open" environment, that is, one under the regulation of an operating system (OS, or "mask") integrated in a microcontroller. This feature makes it possible to meet certain specifications for high-efficiency multi-application cards, combing ticketing and electronic wallet functions, for example. MIFARE Pro offers 20 kB of ROM, 256 bytes of RAM, and 8 k or 2 kB of E2PROM.

— MIFARE Pro X, Smart MX, DESFire:
The members of this class supplement the functions of MIFARE Pro with a "Famex" mathematics coprocessor to facilitate and accelerate the calculation of public-key algorithms (RSA, elliptic curves, etc.), and have extended memory resources to meet the requirements of interbank electronic wallets, passports and visas.

Figure 6.1 shows in a highly summarized form the main characteristics of the members of this class. In view of all this, for the purposes of correct operation, the main characteristic of these circuits is that they require the same base station concept and performance (in terms of power, antenna, etc.), thus greatly facilitating the design and production of a system. This is the main reason for this choice in relation to the following examples.

The next paragraphs will examine various cases of applications:

— for example, when the mechanical size of the antenna is specified;

— when the antenna size is to be optimized,

— and so on.

	contact ISO 7816	contactless ISO 14443 part A	μC	Contact clock MHz	Contactless clock MHz	ROM K byte	RAM byte	E2PROM K byte	encryption	DES	MMU
Mifare UltraLight	no	yes	-	-	fc	-	-	0.512	-	-	-
Mifare Light	no	yes	-	-	fc	-	-	0.512	crypto 1	-	-
Mifare standard	no	yes	-	-	fc	-	-	1	crypto 1	-	-
Mifare extended	no	yes	-	-	fc	-	-	4	crypto 1	-	-
Mifare Pro	yes	yes	8051	1 to 5	fc / 4	32 to 48	1280	4 to 8	Famex	3 × DES	-
Mifare Pro X	yes	yes	8051	1 to 8	fc	48 to 64	2304	8 to 16	Famex 4096 bit	3 × DES	yes

Figure 6.1 MIFARE class circuits conforming to standard 14 443 – A

6.2.2 The base station

We shall start with the specification of the base station.

First case of application – with a specified antenna radius Let us assume, for simple mechanical design reasons, that the overall dimensions of the base station antenna have been specified. By way of example, let us examine the case in which the antenna is assumed to be circular, with a diameter of 10 cm.

Characteristics of the base station antenna This section describes the principal mechanical and electrical characteristics used for the design of the base station antenna.

Mechanical characteristics

Number of turns:

$$N_1 = 1 \quad \text{(making the calculations so much more straightforward!)}$$

Geometry of the circular turn(s):

$$\text{diameter } D_1 = 10 \text{ cm}$$

$$\text{radius } r_1 = 5 \text{ cm}$$

Note: If the antenna is square $(a \cdot a)$ or rectangular $(a \cdot b)$, then, as a first approximation, we can calculate the equivalent radius of a circular antenna $r_{1\,equi} = 2\pi(a \cdot b)$.

Length of one turn:
$$l_1 = 2\pi r_1$$
$$= 2 \times 3.14 \times 0.05$$
$$= 0.314 \text{ m}$$

Surface area of one turn of the antenna:
$$s_1 = \pi \frac{D_1{}^2}{4}$$
$$= 0.314 \times \frac{0.1^2}{4}$$
$$= 78.5 \times 10^{-4} \text{ m}^2$$

Total surface area of the antenna:
$$S_1 = N_1 s_1$$
$$S_1{}^2 = (N_1 s_1)^2$$
$$= (1 \times 78.5 \times 10^{-4})^2$$
$$= 6\,162.25 \times 10^{-8} \text{ m}^2$$

Radiation characteristics of the antenna

Radiation resistance of the antenna:

$$Ra = \frac{31\,200}{\lambda c^4} \cdot (N_1 s_1)^2$$

$$= \text{const} \cdot S_1{}^2$$

$$= 0.13 \times (78.5 \times 10^{-4})^2$$

$$= 8 \,\mu\Omega$$

Electrical parameters of the physical circuit of the base station antenna If the mechanical dimensions of the base station antenna have been specified, the electrical parameters of the base station antenna can be estimated by means of some theoretical calculations. You will find all the necessary information in Chapter 8, which deals with antennae.

At the end of this process, it is always instructive to measure the electrical parameters of the antenna as constructed (which will certainly be very close to the theoretical values ... but never quite the same!).

Examples of measured values:

$$\text{inductance } L_1s = 500 \,\text{nH}$$

$$\text{series resistance } R_1s = 0.5 \,\Omega$$

$$\text{parasitic parallel capacitance } C_1p = 30 \,\text{pF}$$

The equivalent circuit of the antenna can then be drawn. It is shown in *Figure 6.2a*.

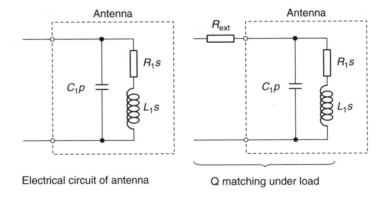

Electrical circuit of antenna   Q matching under load

Figure 6.2a Series circuit of the base station antenna

Intrinsic resonant frequency of the base station antenna For information, let us calculate the (series) intrinsic resonant frequency *fs* of the base station antenna.

Assuming that $\omega s = 2\pi fs$, *fs* is such that

$$L_1 s C_1 p \cdot \omega s^2 = 1$$

that is,

$$\omega s^2 = \frac{1}{(30 \times 10^{-12} \times 500 \times 10^{-9}) \times 10^{18}}$$

$$\omega s = 0.258 \times 10^9 \,\text{rd/s}$$

that is,

$$fs = 41.11 \, \text{MHz}$$

Impedance of the base station antenna inductance At the specified operating frequency of the carrier fc, which is 13.56 MHz, the inductance of the base station antenna has an impedance of

$$
\begin{aligned}
Z_{L_1 s} &= L_1 s \cdot \omega \\
&= 500 \times 10^{-9} \times 85.157 \times 10^6 \\
&= 42.56 \, \Omega
\end{aligned}
$$

Intrinsic quality factor of the base station antenna At this frequency, the antenna constructed in this way (in its series or parallel form) has an intrinsic quality factor, whose value is given by the formula

$$Q_1 s = \frac{L_1 s \cdot \omega}{R_1 s} = \frac{45.56}{0.5} = 85.157$$

No comment required.

Application-specific parameters of the base station antenna To meet the requirements of the proposed application, given its bit rate and associated temporal characteristics (particularly the pause time $Tp = 3 \, \mu s$ of the 100% ASK modulation), we must adapt (in fact, reduce) the quality factor of the circuit. In the case of the Mifare class, according to ISO 14 443 (Part 2A in this case), as shown in Chapter 3, we must bring Q_1 under load at a maximum value of approximately $Q_{1\,\text{load}} = 35$, which will subsequently be denoted Q_1 for simplicity.

To decrease the value of Q_1 under load at 13.56 MHz, we must then connect in series with the antenna an additional resistance R_{ext}. To calculate the value of R_{ext}, assuming that the value of $C_1 p$ is negligible, we can write

$$
\begin{aligned}
Q_1 &= 35 \\
&= \frac{L_1 s \cdot \omega}{R_1 s + R_{\text{ext}}}
\end{aligned}
$$

and therefore

$$
\begin{aligned}
R_{\text{ext}} &= \frac{L_1 s \cdot \omega}{Q_1} - R_1 s \\
&= \frac{45.56}{35} - 0.5 \\
&= 0.716 \, \Omega
\end{aligned}
$$

Parallel equivalent circuit of the physical antenna of the base station Using a conventional series/parallel transformation (see appendices), we obtain the parallel

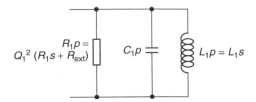

Figure 6.2b   Parallel circuit of the base station antenna

equivalent circuit of the whole system represented by the base station antenna, in the form of a conventional anti-resonant circuit (*Figure 6.2b*).

$$L_1 p = L_1 s = 500\,\text{nH}$$

$$R_1 p = Q_1{}^2(R_1 s + R_{\text{ext}}) = 35^2(0.5 + 0.716)$$

$$= 1496\,\Omega$$

$$C_1 p = 30\,\text{pF}$$

Clearly, the series intrinsic resonance is equal to the parallel intrinsic resonance, that is, 41.11 MHz.

Matching the source impedance (amplifier) and load impedance (base station antenna circuit)

The global impedance of the anti-resonant circuit produced in this way ($R_1 p = 1496\,\Omega$ or alternatively $R_1 s' = (R_1 p)/Q_1{}^2 = 1.216\,\Omega$) is not matched in any way (in terms of power) to the output impedance (real only) of the power amplifier that drives the antenna.

Generally, at 13.56 MHz the output impedance of the amplifier is set at $R_{\text{out}} = 50\,\Omega$. One of the main reasons for this choice is the frequent use of connecting cables that have this characteristic impedance.

This impedance matching is easy to achieve, particularly by using a capacitance bridge as shown in *Figure 6.3* (see Chapter 3 again if necessary).

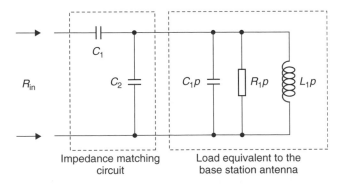

Impedance matching   Load equivalent to the
circuit   base station antenna

Figure 6.3   Capacitance bridge for impedance matching

As demonstrated previously, the matching conditions are satisfied when the values of the two capacitances are practically equal to

$$C_1 = \frac{1}{\omega\sqrt{R_{out}Rp}}$$

$$C_2 = \frac{1}{Lp \cdot \omega^2} - C_1 - Cp$$

and therefore, in the case of our example,

$$C_1 = \frac{1}{85.156 \times 10^6\sqrt{50 \times 1\,496}}$$

$$= 43\,\text{pF}$$

$$C_2 = \frac{1}{500 \times 10^{-9}(85.156 \times 10^6)^2} - 43 - 30$$

$$= 202.75\,\text{pF}$$

At this stage, all the values of the components of the electrical circuit have been determined.

Now we must consider the various electrical stresses (voltage, current, power) to which these components will be subjected.

Electrical stresses on each element   Once all the electrical values of the various elements are known, we must estimate the electrical stresses (current/voltage/power) that all these elements must withstand.

Electrical parameters of the amplifier (integrated circuit) driving the base station antenna

NOTE

The following parameters have been taken from the specification of the Philips Semiconductors MF RC 500 integrated circuit for 13.56 MHz proximity base station applications.

— Maximum current that the base station integrated circuit can deliver:

$$I_{max} = 200\,\text{mA peak}$$

— Mean maximum current that the base station integrated circuit can withstand:

$$I_{moy\,max} = 2\frac{I_{peak\,ant}}{\pi}$$

$$= \frac{2 \times 0.2}{3.14}$$

$$= 127\,\text{mA}$$

— Maximum dissipation possible for the integrated circuit:

$$V_{dd}\,I_{moy\,max} = 5 \times 127 \times 10^{-3}$$

$$= 0.635\,\text{W}$$

NOTE

The power in watts supplied by the integrated circuit to its load is conserved.

The electrical stresses

Assuming that the maximum power to be delivered by the amplifier to its load is known, for example,

$$P_1 = 0.6\,\text{W rms}$$

then

— **Input impedance:**

$$R_{\text{in}} = R_{\text{out}} = 50\,\Omega$$

— **Input current:**

$$I_{\text{in rms}} = \sqrt{\frac{P_1}{Z_{\text{in}}}} = 0.109\,\text{A rms}$$

— **Parameters relating to C_1:**
Current flowing through C_1:

$$I_{C_1} = I_{\text{in}}$$
$$= 0.109\,\text{A rms}$$

Voltage across the terminals of C_1:

$$U_{C_1} = \frac{1}{C_1 \cdot \omega} I_{C_1}$$
$$= \frac{1}{43 \times 10^{-12} \times 85.157 \times 10^6} 0.109$$
$$= 27.3\,\text{V rms}$$

— **Antenna parameters:**
Current in the antenna:

$$I_{\text{ant}} = I_{L_1 s} = \sqrt{(P_1 Q_1)(L_1 s \cdot \omega)}$$
$$= \sqrt{(0.6 \times 35)(500 \times 10^{-9} \times 85.157 \times 10^6)}$$
$$= 0.7\,\text{A rms}$$

Voltage across the terminals of L (antenna):

$$U_{L_1 s} = (L_1 s \cdot \omega) \cdot I_{L_1 s}$$
$$= 42.56 \times 0.7$$
$$= 29.89\,\text{V rms}$$

$$U_{L_1 s\,pp} = 2 \times 29.89 \times 1.414$$

$$= 84.53 \, \text{V pp}$$

— **Parameters relating to R_{ext}:**
Current flowing in R_{ext}:

$$I_{R_{ext}} = I_{L_1 s}$$

Power dissipated in R_{ext}:

$$P_{R_{ext}} = R_{ext} I_{L_1 s}{}^2$$

Values of magnetic flux, fields and inductions   So far, we have only determined the electrical values, in the form of volts, amperes, power, and so on. In the design of a contactless system, in order to ensure its correct operation we must obviously examine and determine all the relevant magnetic parameters such as flux, fields and induction that come into play in this type of application.

Sequence of operations   Let us start by examining the following sequence of operations:

— if we know the value of the inductance $L_1 s$ of the base station antenna, we can calculate its impedance at the operating frequency of the system;

— the voltage V_1 developed across its terminals enables us to deduce the value I_1 of the current flowing through it;

— the value I_1 of the current flowing through the antenna inductance of the base station enables us to determine

— the value of the flux Φ_1 produced by the base station antenna,

— the strength of the magnetic field H_0 produced at the centre of the base station antenna,

— the value of the magnetic induction B_0 at the centre of the base station antenna, and, consequently, the values of the remote magnetic fields and induction.

Summary:

$$\boxed{L_1 s \Rightarrow Z_{L_1 s} \Rightarrow \frac{V_1}{Z_{L_1 s}} \Rightarrow I_1 \Rightarrow \Phi_1 \Rightarrow L_1 s \cdot I_1 \Rightarrow H_0 \Rightarrow B_0}$$

Magnetic flux produced at the centre of the base station antenna   The flux produced at the centre of the antenna, Φ_0, is caused by the flow of the current $I_{L_1 s}$ in the inductance $L_1 s$ of the antenna. Its value is

$$\Phi_0 = L_1 s \cdot I_{L_1 s}$$

$$= 0.5 \times 10^{-6} \times 0.7$$

$$= 350 \, \text{n Wb/m}^2 \, \text{rms}$$

Magnetic field produced at the centre of the base station antenna   The flow of the current $I_{L_1 s}$ in the N_1 turns, which makes up the base station antenna, also creates a magnetic field $H_{0\,cm}$ in the centre of the antenna, whose equation, for a flat circular antenna, is written as

$$H_0 = \frac{N_1 I_{L_1 s}}{2r_1}$$

$$= \frac{1 \times 0.7}{2 \times 0.05}$$

$$= 7\,\text{A/m rms}$$

Magnetic induction in the centre of the base station antenna   The above magnetic field H_0 causes the associated appearance of an induction B_0 in the centre of the base station antenna. In the case of a medium such as a vacuum or air,

$$B_0 = \mu_0 H_0$$

$$= 12.56 \times 10^{-7} \times 7$$

$$= 8.79\,\mu\text{T rms}$$

NOTES

— If the magnetic medium in which the base station antenna is located is not air, it will be necessary to correct the value of μ_0 by $\mu = \mu_0 \cdot \mu_r$ to allow for the magnetic environment.

— For information, you should note that the induction created in air at a distance $d = 2r_1 = 10\,\text{cm}$ along the axis of the antenna having a radius $r_1 = 5$ cm is as follows (refer back to Chapter 1 if necessary):

$$B_{10} = \frac{\mu_0 H_0}{11.8}$$

$$= \frac{12.56 \times 10^{-7} \times 7}{11.8}$$

$$= 760\,\text{nT}$$

Power radiated by the base station antenna   The flow of current in the antenna produces radiation from the antenna, the power of this radiation, EIRP (Equivalent Isotropic Radiated Power), being expressed by the following equation:

$$\text{power radiated by the antenna } Pa = Ra \cdot I_{L_1 s}^2$$

$$= 8 \times 10^{-6} \times 0.7^2$$

$$= 3.92\,\mu\text{W}$$

The following insert and *Figure 6.4* summarize the key parameters of this first proximity application.

– specific assumptions for the planned application:

$L_1 s = 0.5 \, \mu\text{H}$ $\qquad N_1 = 1$

$R_1 s = 0.5 \, \Omega$ $\qquad r_1 = 5 \, \text{cm}$

$R_{\text{ext}} = 0.7 \, \Omega$ $\qquad Q_1 = 35$ under load

$C_1 p = 30 \, \text{pF}$ $\qquad P = 0.6 \, \text{W}$

– performance of the base station:

$I_{\text{ant}} = 0.7 \, \text{A}$ $\qquad B_0 = 8.79 \, \mu\text{T}$

$V_{\text{ant}} = 30 \, \text{V}$ $\qquad B_{10} = 760 \, \text{nT}$

$C_1 = 43 \, \text{pF}$ $\qquad H_{10} = H_{2r} = \dfrac{H_{10}}{11.8} = 0.6 \, \text{A/m}$

$C_2 = 203 \, \text{pF}$ $\qquad Pa = 3.9 \, \mu\text{W}$

$H_0 = 7 \, \text{A/m}$ $\qquad P = 0.6 \, \text{W}$ plus the power dissipated in the external resistance R_{ext}

$N_1 = 1$ $\qquad s_1 = 78.5 \, \text{cm}^2$

$L_1 = 500 \, \text{nH}$ $\qquad Ra = 8 \, \mu\Omega$

$R_1 = 0.5 \, \Omega$ $\qquad Pa = 3.92 \, \mu\text{W}$

$Q_1 = 35$ $\qquad fc = 13.56 \, \text{MHz}$

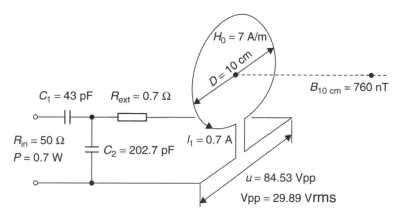

Figure 6.4 Summary of the first proximity application

Second example of application – with a specified maximum radiated power   Moving
away from the previous example, in which the mechanical dimensions of the antenna
were specified at the outset, we shall now examine the problem from a completely dif-
ferent viewpoint, by finding and optimizing the components for a case where nothing is
specified in advance concerning the design of the base station components, except for
the radiated power, which is as follows:

radiated power = specified, or known, or constant, or required
to conform to a standard value

Let us begin with a few theoretical matters. We know, by definition, that the radiated
power is as follows:

$$Pa = Ra \cdot I_{ant}^2$$
$$= (\text{const} \cdot S_1^2) \cdot I_{ant}^2$$
$$= \text{const} \cdot [N_1(\pi r_1)^2]^2 \cdot I_{ant}^2$$

The magnetic field in the centre of the antenna, H_0, is as follows:

$$H_0 = \frac{N_1 I_{ant}}{2r_1}$$

that is,

$$r_1 = \frac{N_1 I_{ant}}{2H_0}$$

Now let us introduce the value of r_1 into the radiated power equation. We find

$$Pa = (\text{const} \cdot N_1^2 \cdot \pi^2 r_1^4) \cdot I_{ant}^2$$
$$= (\text{const} \cdot N_1^2 \cdot \pi^2 N_1^4 I_{ant}^4) \cdot \frac{I_{ant}^2}{(2H_0)^4}$$
$$= \frac{\text{const} \cdot \pi^2 N_1^6 I_{ant}^6}{(2H_0)^4}$$

and therefore

$$Pa = \frac{\text{const} \cdot \pi^2 N_1^6}{(2H_0)^4} I_{ant}^6$$

Also, for the base station antenna circuit, we have shown in Chapter 3 that the
equation relating the current in the antenna and the power in watts P_1 dissipated in the
load R_1s (where the tuned circuit is resonating) takes the form

$$I_{ant} = \sqrt{\frac{P_1 Q_1}{L_1 \cdot \omega c}}$$

By transferring the value of I_{ant} of the above equation, we obtain

$$Pa = \frac{\text{const} \cdot N_1^6 \cdot \pi^2}{(2H_0)^4} \cdot \left(\frac{P_1 Q_1}{L_1 \cdot \omega c}\right)^3 = f(P, H_0, Q_1, L_1, N_1)$$

This set of equations enables us to estimate, for a given system (N_1, L_1, Q_1, H_0), on the one hand the power in watts that must be supplied as a function of the desired radiated power, that is,

$$P_1 = \frac{L_1 \cdot \omega c}{N_1^2 Q_1} \cdot \sqrt[3]{\frac{Pa \cdot (2H_0)^4}{\text{const} \cdot \pi^2}} = f(Pa)$$

and on the other hand the associated value of the antenna current that produces the magnetic field required by the application, that is,

$$I_{ant} = \sqrt[6]{\frac{Pa \cdot (2H_0)^4}{\text{const} \cdot \pi^2 N_1^6}} = f(Pa)$$

A few interesting comments Let us return for a moment to the expression of $Pa = f(P, H_0, Q_1, L_1, N_1)$ which we found in the previous section, and let us start by reordering the terms of this equation; thus:

$$Pa = P^3 \left[\frac{\text{const} \cdot N_1^6 \cdot \pi^2}{(2H_0)^4} \cdot \frac{Q_1^3}{(L_1 \cdot \omega c)^3}\right] = f(P, H_0, Q_1, L_1, N_1)$$

Everything else being equal, $Pa = f(P^3)$.

We shall now look at a numbered example of this equation.

Let us assume that the base station amplifier can deliver a power varying from $P_1 = 0.5\,\text{W}$ to $P_2 = 4\,\text{W}$ (warning: this is only an example; in reality, the power of an amplifier would be more likely to decrease from 4 W to 0.5 W).

From the above formula we can deduce that, if the power in watts P, supplied to the base station antenna, varies with a ratio $P_2/P_1 = 4/0.5 = 8$, the radiated power Pa varies proportionally to the cube of the ratio of P, that is, with a ratio of $8^3 = 512$.

Thus if, for a value of P_1 of 0.5 W supplied to a given circuit, we obtain a radiated power of $Pa = 20\,\mu\text{W}$, then, for a supplied power $P_2 = 4\,\text{W}$, we will obtain $Pa = 20\,\mu\text{W} \times 512$, and therefore, in approximate terms, $Pa_2 \approx 10\,\text{mW}$.

NOTE

The conclusion of this note usually surprises readers who are already familiar with contactless applications. The key to the mystery lies in "everything else being equal" at the start of the section.

Let us return for a moment to the content of the formula and the way in which we increase the power P.

It is easy to see that, for a fixed and determined system, the values and coefficients such as frequency, number of turns, L_1 and Q_1, are totally independent of the variation in power. In this configuration, obviously, whenever P increases, the current flowing in the antenna and therefore the magnetic field H_0 also increases. If we want to retain the initial assumption of our argument, we must find a way of keeping H_0 constant. This is not a simple matter, unless, while keeping the same number of turns, the same value of L_1 and the same Q_1, we change the mechanical shape and/or size of the antenna. This is equivalent to changing its radiation resistance Ra.

Another view of the same problem   No, I am not trying to make you suffer, but sometimes we need to investigate every aspect of a problem. So let us approach it from another direction. We know that

$$Pa = Ra \cdot I_{ant}^2$$

and

$$I_{ant} = \sqrt{\frac{P_1 Q_1}{L_1 \cdot \omega}}$$

and, transferring the value of I_{ant}^2 into the first equation,

$$Pa = Ra \cdot \frac{P_1 Q_1}{L_1 \cdot \omega} = f(P_1)$$

Some more very interesting comments   Once again, if everything else is equal (L_1, Q_1, and, to stress it this time, with the same Ra), if P_1 increases from 0.5 W to 4 W, in other words to a value eight times greater, and if the initial value of Pa was $20\,\mu W$, then Pa will change to $8 \times 20 = 160\,\mu W$. This is the usual situation with standard applications and most designers. The physical structure of the antenna is left as it is, and if the range is insufficient, more power is added.

For the same L_1 and Q_1 and with the same ratio of power P supplied by the power stage (from 0.5 W to 4 W, i.e. a ratio of 8), if, at the same time, the radiated power Pa changed to 10 mW, this would mean that the radiation resistance Ra had simultaneously increased from the ratio $(10\,mW)/(160\,\mu W)$, that is, a ratio of $10\,000/160$.

Given that the value of Ra is proportional to the square of the surface area of the antenna, if we assume that the antenna technology remains the same (with just one turn), this would mean that the square of the surface areas $(S_2/S_1)^2$ had also increased in the ratio of $10\,000/160 = 62.5$ in other words that the ratio of the surface areas S_2/S_1 had increased in the ratio of the square root of $10\,000/160$, that is, 7.9.

Thus, if the initial diameter of the antenna coil was $dia_1 = 15.8\,cm$,

$$r_1 = 7.9\,cm$$

$$S_1 = 196.23\,cm^2$$

the new surface area would be as follows:

$$S_2 = 7.9 \times 196.23$$

$$= 1\,551.34\,cm^2$$

that is, a value of

$$r_2 = 22.23\,cm$$

and
$$dia_2 = 44.46 \, cm$$

To summarize, for the same value of inductance L_1 ($L_1 = 500 \, \mu H$) and Q_1 ($Q_1 = 35$):

$$
\begin{array}{lll}
dia_1 = 15.8 \, cm & dia_1 = 15.8 \, cm & dia_2 = 44.46 \, cm \\
P = 0.5 \, W & P = 4 \, W & P = 4 \, W \\
Pa = 20 \, \mu W & Pa = 160 \, \mu W & Pa = 10 \, \mu W
\end{array}
$$

Example

Let us assume, for example, that

— in order to comply with the current regulations, we wish to limit the radiated power of the base station antenna – Pa – (the maximum permitted value, for example) to
$$Pa = 20 \, \mu W \text{ rms}$$

— and that, additionally, in order to make a given transponder operate at a distance of 10 cm, we need a magnetic induction of $B_0 = 5 \, \mu T$ rms in the centre of a base station antenna, still consisting of one turn ($N_1 = 1$).

We start by calculating H_0:

$$
\begin{aligned}
H_0 &= \frac{B_0}{\mu_0} \\
&= \frac{5 \times 10^{-6}}{4\pi \times 10^{-7}} \\
&= 4 \, A/m \, rms
\end{aligned}
$$

enabling us to calculate I_{ant}:

$$I_{ant}{}^6 = \frac{Pa \, (2 \times H_0)^4}{const \cdot \pi^2 N_1{}^6}$$

When const $= 0.13$, we obtain

$$I_{ant} = 630 \, mA \, rms$$

We can then calculate the radius of the turn r_1, which gives us

$$r_1 = \frac{N_1 I_{ant}}{2 \times H_0}$$

and therefore

$$r_1 = 7.9 \, cm$$

and

$$D_1 = 2r_1 = 15.8 \, cm$$

For information, the magnetic induction B created at a distance $d = 2 \times r_1 = 15.8$ cm is then

$$\Rightarrow B_{15.8} = \frac{B_0}{11.8} = \frac{5000}{11.8}$$
$$= 428 \, \text{nT rms}$$

Let us make a quick check of this solution:

surface area of one turn:

$$s_1 = \pi r_1^2 = 3.14 \times 0.079^2$$
$$= 196.23 \, \text{cm}^2$$
$$N_1 = 1$$
$$S_1 = N_1 s_1$$

and therefore

$$Pa = \text{const} \cdot S_1^2 I_1^2$$
$$= 0.13 \times (196.23 \times 10^{-4})^2 \times 0.63^2$$
$$= 20 \, \mu\text{W} \quad \text{Q.E.D.}$$

Now let us assume that, in this new example, we have decided to use the same electrical values and parameters of the antenna as those specified in the first example (but be careful, the radius r_1 of the base station antenna has been modified), so that now

$2r_1 = 15.8 \, \text{cm}$

$L_1 s = 500 \, \text{nH}$

$R_1 s = 0.5 \, \Omega$

$Q_1 = 35$

and therefore

$$Rp = Q_1^2 (R_1 s + R_{\text{ext}})$$
$$= 1496 \, \Omega$$

The voltage developed across the terminals of the base station antenna then becomes

$$U_{L_1 s} = Z_{L_1 s} I_{\text{ant}}$$
$$= Ls \cdot \omega \cdot I_{\text{ant}}$$
$$= (500 \times 10^{-9} \times 85.157 \times 10^6) \times 0.630$$
$$= 26.8 \, \text{V rms}$$

or $= 75.7$ V pp in place of the 84.53 V pp found in the previous example.

Similarly, the magnetic flux produced by the base station antenna in this case will be

$$\Phi_{ant} = L_S I_{ant}$$
$$= 0.5 \times 10^{-6} \times 0.630$$
$$= 315 \text{ nWb/m}^2$$

in place of the 350 nWb/m^2 calculated in the previous example.

If, for example,

$$Pa = 20 \,\mu\text{W rms}$$

$$\text{const} = 0.13$$

$$Q_1 = 35$$

$$H_0 = 4 \text{ A/m}$$

$$L_1 s \cdot \omega c = 42.56 \,\Omega$$

$$N_1 = 1$$

and therefore

$$P_1 = 0.48 \text{ W rms}$$

Check

If the power dissipated in the load is $P_1 = 0.48$ W, this power is dissipated in the resistance Rp connected in parallel with the $L_1 s$, C, circuit of the base station.

Given that

$$P_1 = R_1 p \cdot Ip^2$$
$$Ip^2 = \frac{P_1}{R_1 p} = \frac{0.48}{1500}$$
$$= 320 \times 10^{-6}$$

and therefore

$$Ip = 17.9 \text{ mA rms}$$

The current flowing in the base station antenna coil is as follows:

$$I_{ant} = Q_1 I_p$$
$$= 35 \times 17.9 \times 10^{-3}$$
$$= 626 \text{ mA} = 0.626 \text{ A rms} \quad \text{Q.E.D.}$$

NOTE

In fact, the power delivered by the amplifier is always (slightly) greater than the value calculated above, since the resistance R_{ext} also dissipates a small amount of power. In the context described above, this supplementary power is

$$P_{R_{ext}} = R_{ext} I_{ant}^2$$
$$= 0.716 \times 0.630^2$$
$$= 0.28\,\text{W}$$

that is,

$$P_{total} = 0.48 + 0.28$$
$$= 0.76\,\text{W}$$

Success!   Returning to the expression giving the value of the power $Pa = Ra \cdot I_{ant}^2$ radiated by the antenna and the value of the magnetic field in its centre, $H_0 = (N_1 I_{ant}^2)/(2r_1)$, we obtain

$$Pa = Ra \cdot I_{ant}^2$$
$$= (\text{const} \cdot S_1^2) \cdot I_{ant}^2$$
$$= [\text{const} \cdot (N_1^2 \cdot \pi r_1^2)^2] \cdot I_{ant}^2$$

and

$$I_{ant} = \frac{H_0 \times 2r_1}{N_1}$$

that is,

$$\boxed{Pa = Ra \cdot \left(\frac{H_0 \times 2r_1}{N_1}\right)^2}$$

an expression which leads to the same result as before.

The following insert and *Figure 6.5* summarize the key parameters of this example.

– **assumptions:**

$Pa = 20\,\mu\text{W}$	$R_1 = 0.5\,\Omega$	$R_{ext} = 0.7\,\Omega$	$N_1 = 1$
$B_0 = 5\,\mu\text{T}$	$L_1 = 0.5\,\mu\text{H}$	$C_1 p = 30\,\text{pF}$	$Q_1 = 35$

– **performance:**

$dia = 15.8\,\text{cm}$ $\qquad\qquad\qquad H_{10} = \dfrac{H_0}{4.19} = 0.95\,\text{A/m}$

$r_1 = 7.9\,\text{cm}$ $\qquad\qquad\qquad B_{10} = \dfrac{B_0}{4.19} = 1.2\,\mu\text{T}$

$H_0 = 4\,\text{A/m}$ $\qquad\qquad\qquad B_{15.8} = B_{2r} = \dfrac{B_0}{11.8} = 428\,\text{nT}$

$d = 10\,\text{cm}$ $\qquad\qquad\qquad I_{ant} = 0.630\,\text{A rms}$

$a = \dfrac{d}{r} = 1.265$ $\qquad\qquad\quad V_{ant} = 75.8\,\text{V pp}$

$(1 + a^2)^{3/2} = 4.19$ $\qquad\qquad P_{tot} = 0.48\,\text{W}\ (+0.26\,\text{W in the external resistance})$

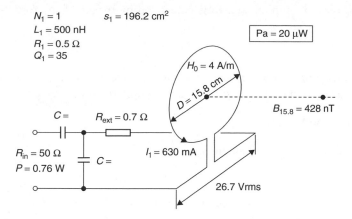

Figure 6.5 Summary of the second proximity application (for a given radiated power)

6.2.3 The transponder

We have nearly finished with the base station. We know the precise magnetic field and induction values that it creates in its centre and therefore as a function of the distance. Now we will look at the application as seen from the transponder, to ensure that this component is compatible with the base station's performance.

The integrated circuit As mentioned at the beginning of the chapter, we will use the Philips Semiconductors Mifare 1 integrated circuit to illustrate these examples.

The document describing its principal electrical characteristics mentions some specific values:

$$C_{\text{cic}} = 16.2\,\text{pF nominal}$$

$$C_{\text{coil}} = 5\,\text{pF}$$

$$C_{\text{parasite}} = 5\,\text{pF}$$

$$C_{\text{mounting}} = 2\,\text{pF}$$

that is,

$$C_{\text{total}} = 28.2\,\text{pF}$$

$$P_{\text{min max}} = 2\,\text{mW}$$

$$V_{\text{ic max}} = 4\,\text{V rms}$$

and therefore

$$R_{\text{ic}} = 8\,\text{k}\Omega \text{ at } V_{\text{ic}} = 4\,\text{V}$$

Important notes about the above values

— The values of C_{ic} and R_{ic} are measured with a network analyser (an Agilent or Marconi model, for example) for a sinusoidal alternating voltage $V_{\text{ic}} = 4\,\text{V rms}$.

—   The power consumption shown is calculated by means of the formula

$$P_{ic} = \frac{V_{ic}{}^2}{R_{ic}}.$$

The transponder antenna and the antenna circuit   The manufacturer of the integrated circuit also provides the following recommendations for use:

$30 < Q_{2}s$ at 13.56 MHz < 60

inductance of the transponder antenna:

$$L_2 < 4.2 \,\mu\text{H}$$

$$L_2 \text{ recommended} = 3.6 \,\mu\text{H}$$

total surface area S_2 of the transponder antenna:

$$S_2 > 133 \,\text{cm}^2$$

where

$N_2 =$ number of turns

$s_2 = \dfrac{S_2}{N_2}$

s_2 max recommended $> 27.785 \,\text{cm}^2$

Example

By way of example, consider the production of a contactless smart card in ISO card format that has to conform to ISO 14 443 ("proximity" applications; in other words, approximately, $d = 10$ to $12\,\text{cm}$).

Given the available mechanical surface area of the card ($85.5 \times 54\,\text{mm}$ in the ISO 7810 ID1 format) (see *Figure 6.6*), we decide to use the largest possible surface area for the transponder antenna, in other words a rectangle measuring $8 \times 4.8\,\text{cm}$, giving us:

chosen antenna surface area:

$$s_{2\,\text{chosen}} = (8 \times 4.8) \,\text{cm}^2 = 38.4 \,\text{cm}^2$$

that is,

$$\text{length of turn} = 2 \times (8 + 4.8) \,\text{cm} = 25.6 \,\text{cm}$$

If we wish to provide the recommended total active surface area, it must be equal to or greater than $133\,\text{cm}^2$. Having adopted this minimum total value, we can then determine the minimum number of turns for the transponder antenna as follows:

$$\text{minimum number of turns, } N_2 = \frac{133}{38.4}$$

$N_{2\,\text{min calculated}} = 3.46$

$N_{2\,\text{chosen}} = 4$

Figure 6.6   Example of a contactless smart card in standard ISO ID1 format

For physical and mechanical reasons, we cannot implement a supplementary resistance or capacitance in the thickness of the card, so we must now estimate the necessary values of the antenna's resistance and quality factor, as well as its intrinsic resonant frequency.

For the Philips Semiconductors Mifare integrated circuits operating at a bit rate of 106 kb/s with a pause time Tp of $3\,\mu s$ for 100% ASK modulation, the value of Q_2s at 13.56 MHz must be of the order of 30 to 60 in normal operation. Let us choose the least favourable value, $Q_2s = 30$, and also select the recommended inductance value $L_2s = 3.6\,\mu H$. This means that

$$R_2 s_{\text{total max}} = \frac{L_2 s \cdot \omega}{Q_2 s}$$

$$= \frac{3.6 \times 10^{-6} \times 85.157 \times 10^6}{30}$$

$$= 10.22\,\Omega$$

$$R_2 p_{\min} = R_2 s \cdot Q_2 s^2$$

$$= 10.22 \times 30 \times 30$$

$$= 9199\,\Omega$$

$$\omega s = \frac{1}{\sqrt{L_2 s \cdot C_{\text{total}}}}$$

$$= \frac{1}{\sqrt{3.6 \times 10^{-6} \times 28.2 \times 10^{-12}}}$$

$$= 99 \times 10^6$$

$$f s_{\text{tag intrins}} = \frac{\omega s}{2\pi}$$

$$= \frac{99 \times 10^6}{6.28}$$

$$= 15.75\,\text{MHz}$$

$$= 1.15 \times 13.56\,\text{MHz}$$

$$= 1.15 fc$$

NOTE

If the operating frequency of the base station is 13.56 MHz, the fact that $f\,s_{\text{tag intrins}}$ is 15.75 MHz is not a violation of the principle, or an error, since many applications are designed for the situation in which more than one card is placed simultaneously in a "stack" in a card holder. However, the possibility of detuning between cards due to mutual induction must be allowed for (refer back to the end of Chapter 2 if necessary).

As mentioned in Chapter 3, we can draw an equivalent series circuit for the antenna. This consists of (see *Figure 6.7a*) $L_2 s$, $R_2 s$ and $C_2 p$ all in parallel with each other.

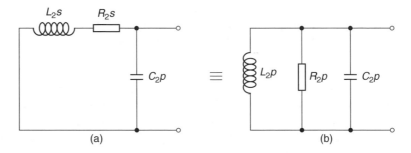

Figure 6.7 Series–parallel conversion of the transponder antenna

As we have just shown, this circuit has its own resonant frequency and quality factor, whose value clearly depends on the frequency at which the antenna operates. This series layout of the antenna circuit can be represented as a parallel equivalent circuit consisting of (see *Figure 6.7b*)

$L_2 p = L_2 s$

$R_2 p = Q_2 s^2 \cdot R_2 s$

$C_2 p$

whose intrinsic resonant frequency is the same as before (naturally) and whose quality factor can now be written as

$$Q_2 p = \frac{R_2 p}{L_2 p \cdot \omega}$$

and whose value is the same as before, that is,

$$Q_2 s = Q_2 p \Rightarrow R_2 p = Q_2 s^2 Rsc$$

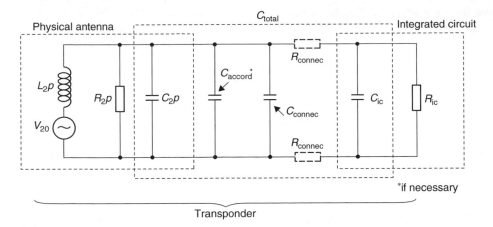

Figure 6.8 Global transponder antenna and load

The transponder antenna circuit/integrated circuit system Now let us look at the complete circuit of the system comprising the antenna and the integrated circuit, shown in parallel form in *Figure 6.8*, in which we can say that

$L_2 p = L_2 s$

$Cp = C_{ic} + C_{con} + C_2 p =$ the sum of:

- the input capacitance of the integrated circuit;

- the parallel capacitances of connections;

- the parallel capacitance of the antenna coil.

$Rp = (R_{ic} R_2 p)/(R_{ic} + R_2 p) =$ the paralleling of the input resistance of the integrated circuit and the equivalent parallel resistance of the antenna coil and

$$L_2 p \cdot Cp \cdot \omega^2 = 1$$

and

$$Qp_2 = \frac{Rp}{L_2 p \cdot \omega}$$

Optimizing the energy transfer between the transponder antenna and the integrated circuit If we want to optimize the energy transfer between the energy source (the transponder antenna) and the integrated circuit, the global impedance of the transponder antenna circuit Zi must be equal to the input impedance of the integrated circuit R_{ic}. As we saw in Chapter 2, if the transponder is tuned to the carrier (which for numerous reasons is not always either completely true or desirable in some cases – see Chapter 3 again), and if $Q_2 s = (L_2 s \cdot \omega)/(R_2 s)$ is large (which is always true, with respect to 1, mainly when we need to optimize the energy transfer), this means that

$$Zi = R_{ic} = L_2 s \cdot \omega \cdot Q_2 s$$

or in other words, in the ideal case,

$$L_2 s = \frac{R_{ic}}{\omega \cdot Q_2 s}$$

$$= \frac{8 \times 10^3}{2 \times 3.14 \times 13.56 \times 10^6 \times 30}$$

$$= 3.13\,\mu\mathrm{H}$$

and therefore a total tuning capacitance Cp such that $L_2 s \cdot Cp \cdot \omega^2 = 1$, that is,

$$Cp = 44\,\mathrm{pF}$$

Clearly, for numerous reasons (deliberate detuning, parasitic connection capacitances, detection of cards arranged in stacks, special collision management procedures, etc.), we have decided not to use these exact values but to adjust the value of L to bring it to the recommended level ($3.6\,\mu\mathrm{H}$) and reset the capacitances accordingly to provide the desired tuning (or detuning).

Having chosen to operate with $Q_2 s = 30$ at 13.56 MHz and having defined the value of $L_2 s$ ($3.6\,\mu\mathrm{H}$), we can now calculate $R_2 p$:

$$R_2 p = Q_2 s \cdot (L_2 s \cdot \omega c)$$

$$= 30 \times (3.6 \times 10^{-6} \times 2 \times 3.14 \times 13.56 \times 10^6) = 9199\,\Omega$$

For the system comprising the transponder antenna and the integrated circuit, we find Rp and Qp_2:

$$Rp = R_2 p // R_{ic}$$

$$= 9199 // 8000 = 4.1\,\mathrm{k\Omega}$$

$$Qp_2 \text{ at } 13.56 = \frac{Rp}{L_2 s \cdot \omega c}$$

$$= \frac{4.1 \times 10^3}{3.6 \times 10^{-6} \times 85.157 \times 10^6} = 13.7$$

NOTE

If we had used the optimal value of $3.13\,\mu\mathrm{H}$, we would have obtained

$$R_2 s_{opt} = \frac{(L_2 s \cdot \omega c)^2}{R_{ic}}$$

$$= \frac{(3.13 \times 10^{-6} \times 85.157 \times 10^6)^2}{8000}$$

$$= 8.88\,\Omega, \text{ in place of } 10.22$$

$$Q_2 s_{opt} = 30$$

$$R_2 p_{opt} = 7992\,\Omega$$

and in this case

$$Qp_{2\,max} = \frac{1}{2}\sqrt{\frac{R_{ic}}{R_2 s}}$$

$$= \frac{1}{2}\sqrt{\frac{8000}{8.88}}$$

$$= 15 \text{ in place of } 13.7.$$

Summary for the base station/transponder system   Assuming that we retain the electrical values of the base station antenna as defined previously, *Table 6.2* summarizes the parameters to be considered for the base station/transponder system.

At this stage in the procedure, the values of, Qp_2, $L_2 s$ and Cp have been determined. Now let us move on to the magnetic values.

Calculating the threshold magnetic field and induction   To calculate these two parameters, we must return to the general equation for the value of the threshold field:

$$H_{dt} = \frac{\sqrt{\left[1 - \left(\frac{\omega c}{\omega r}\right)^2\right]^2 + \left(\frac{L_2 p \cdot \omega c}{Rp}\right)^2}}{\omega \cdot \mu_0 \cdot N_2 s_2} V_{ic\,min\,typ}$$

and

$$B_{dt} = \mu_0 \cdot H_{dt}$$

Examples ... and sub-examples

For the two sets of sub-examples shown below, I suggest that we use an operating carrier frequency $fc = 13.56\,\text{MHz}$ and the Mifare 1 integrated circuit whose operating voltage, as indicated above, is $V_{ic\,min\,typ} = 4\,\text{V}$ (independently of the operating carrier frequency, of course).

(a)   First, let us examine the case where the resonant frequency fr of the transponder is equal to that of the carrier, that is, $fr = fc = 13.56\,\text{MHz}$. In this case,

H_{dt} at 13.56 MHz

$$= \frac{4}{4 \times 3.14 \times 10^{-7} \times 4 \times 38.56 \times 10^{-4} \times 13.7 \times 85.157 \times 10^6}$$

$$= \frac{4}{22.6}$$

$$= 177\,\text{mA/m}$$

and therefore

$$B_{dt} \text{ at } 13.56\,\text{MHz} = 222\,\text{nT}$$

(b)   Now let us look at two cases that differ slightly from the previous case, in which the base station still operates with a carrier frequency of $fc = 13.56\,\text{MHz}$ and $V_{ic\,min} = 4\,\text{V}$ and see what happens when, for the same inductance of the

Table 6.2

The values in the following example are specified for a system operating with a carrier frequency $fc = 13.56$ MHz and a transponder (an ISO format card) tuned to $fs = 15.75$ MHz

$$fc \quad = 13.56\,\text{MHz}$$
$$\omega c \quad = 85.157 \times 10^6 \,\text{rd/s}$$

Base station

$$L_1 \quad = 0.5\,\mu\text{H}$$
$$N_1 \quad = 1$$
$$r_1 \quad = 7.9\,\text{cm}$$
$$I_{\text{ant}} \quad = 630\,\text{mA rms}$$
$$V_{\text{ant}} \quad = 26.77\,\text{V rms}$$
$$H_0 \quad = 4\,\text{A/m rms}$$
$$B_0 \quad = 5\,\mu\text{T rms}$$
$$Pa \quad = 20\,\mu\text{W rms}$$

Transponder

Antenna

$$L_{2}s = L_{2}p \quad = 3.6\,\mu\text{H}$$
$$N_2 \quad = 4\,\text{turns}$$
$$s_2 \quad = 38.56\,\text{cm}^2$$
$$S_2 = N_2 s_2 \quad = 154.24\,\text{cm}^2$$
$$Cp \quad = 28.2\,\text{pF} \qquad \Rightarrow Lsc \times Cp \times \omega^2 = 1 \text{ at } 15.75\,\text{MHz}$$

and therefore $\omega s \quad = 1.16 \times \omega c$

if $\quad Q_{2}s$ at $13.56 = 30$

$$R_{2}p \quad = 9199\,\Omega$$

and therefore $R_{2}s \quad = 10.22\,\Omega$

Mifare 1 integrated circuit

$$V_{\text{ic typ}} \quad = 4\,\text{V rms} \qquad\qquad \text{typical minimum operating voltage}$$
$$R_{\text{ic}} \quad = 8\,\text{k}\Omega \text{ at } 4\,\text{V}$$
$$C_{\text{ic}} \quad = 16.2\,\text{pF}$$
$$P_{\text{typ}} \quad = \frac{V_{\text{ic}}^2}{R_{\text{ic}}} \qquad\qquad \text{to start operation}$$
$$= \frac{4^2}{8 \times 10^3}$$
$$= 2\,\text{mW}$$

System consisting of transponder antenna and integrated circuit

$$Rp \quad = R_{2}p // R_{\text{ic}}$$
$$= 9199 // 8000$$
$$= 4.1\,\text{k}\Omega$$

and therefore Qp_2 at $13.56 = \dfrac{Rp}{L_{2}s \cdot \omega c}$

$$= \frac{4.1 \times 10^3}{L_{2}s \cdot \omega c} = 13.7$$

transponder antenna ($L_2 p = $ const), the tuning frequency of the antenna is modified by changing, for example, the tuning capacitance (e.g. by reducing the parasitic and connection capacitances where necessary):

— let us tune a first transponder to $fr = 16\,\text{MHz}$
  that is, B_{dt} at $16\,\text{MHz} = 825\,\text{nT}$;

— let us tune a second transponder to $fr = 19\,\text{MHz}$
  that is, B_{dt} at $19\,\text{MHz} = 1507\,\text{nT}$.

In other words, the transponder described in the previous sections, tuned to 15.75 MHz, requires a minimum field strength of approximately 825 nT to operate correctly with a base station whose carrier frequency is 13.56 MHz, so that $825/222 = $ approximately four times greater (400% !) compared with a transponder tuned exactly to the carrier.

These numbered examples also enable to estimate, for the same magnetic field H_0 produced by the base station antenna, and for one transponder considered in isolation, the difference in operating range between a transponder exactly tuned to the carrier frequency and a transponder which has been deliberately detuned for operation in a stack, for example (see *Table 6.3*). Clearly, the last observation is applicable to all calculations relating to the effect of tolerances, dispersions, and so on, of the values of the transponder tuning frequency as a function of the tolerances and dispersions of the components.

Table 6.3

Operating frequency of the base station, fc	MHz	13.56	13.56	13.56
Operating frequency of the transponder, ft	**MHz**	**13.56**	**16**	**19**
Ratio ft/fc		1	1.18	1.40
B_0	μT	5	5	5
B_{dt} for $V_{ic\,min} = 4\,\text{V}$	nT	222	825	1507
$B_0/B_{dt} = (1 + a)^{3/2}$	–	22.5	6	3.3
and therefore $a = d/r$	approx.	2.63	1.65	1.25
Given that $r_1 = 7.9\,\text{cm}$: operating distance, d	**cm**	**21**	**12.6**	**6.3**

It's an ill wind that blows nobody any good ...   Why such a peculiar heading for this section?

Suppose, just for fun, that we decide to use a deliberately detuned transponder. For the same base station, as shown in *Table 6.3*, the operating range of this transponder will be smaller than that of an optimized transponder. Hopeless! To illustrate our example, let us say for the sake of simplicity that it now only operates at up to 10 cm.

We shall now put this particular transponder in a stack with other transponders of the same type. After stacking, the phenomenon of mutual induction between all the antennae present causes the whole set to detune, and the "global" tuning frequency of the stack decreases. Hooray!

In fact, although the set of transponders in the magnetic field absorbs more energy (and therefore the operating range of each of them should normally decrease) because the "global tuning" of the stack approaches the carrier frequency, their individual sensitivities increase because the minimum operating threshold H_{dt} is reduced (and therefore the operating range increases)!

In conclusion, as seen by the end user, everything will be just fine for him because one transponder in isolation will operate at 10 cm and the same will apply, for practical purposes, when several transponders are stacked, since each of them is individually detuned at the outset (to a higher frequency than that of the carrier used). Clever, isn't it?

Now that you have some idea of the complexities involved in implementing contactless technology, it is up to you to find some smart solutions to your problems ... preferably without infringing any of the numerous patents!

Voltage induced in the transponder by the base station

To avoid confusing the calculations (or your brain cells) any further, let us continue with our example by assuming, for simplicity, that $L_2 p = 3.6\,\mu$H and that the tuning frequency of the transponder, fr, is equal to the carrier frequency fc, that is, $fs = fc = 13.56$ MHz.

On this assumption, we can now calculate the typical voltage $V_{20\,typ}$ which must be developed in the idle state across the terminals of the transponder antenna. We know that (see Chapter 2 again if necessary)

$$V_{ic\,typ} = \frac{Rp}{L_2 p \cdot \omega} V_{20\,typ}$$

or alternatively

$$V_{ic\,typ} = Qp_2 \cdot V_{20\,typ}$$

To obtain $V_{ic\,typ} = 4$ V rms, the typical induced voltage must be as follows:

$$V_{20\,typ} = V_{ic\,typ}\frac{1}{Qp_2}$$

$$= \frac{4}{13.7}$$

$$= 291\,\text{mV rms}$$

Value of the typical minimum coupling coefficient Continuing with our example, in which the voltage applied to the base station antenna is $V_1 = 26.8$ V rms, we can now estimate the value of $k_{min\,typ}$ required for the application to operate correctly as follows:

$$V_{20\,typ} = V_1 \cdot k_{min\,typ}\sqrt{\frac{L_2}{L_1}}$$

and therefore

$$k_{min\,typ} = \frac{V_{20\,typ}}{V_1\sqrt{\frac{L_2}{L_1}}}$$

$$= \frac{291 \times 10^{-3}}{26.8\sqrt{\dfrac{3.6}{0.5}}}$$

$$= 0.00406$$

and therefore

$$k_{\min\,typ} = 0.4\%$$

Value of the typical minimum operating distance   This (very) small value enables us to calculate the theoretical maximum possible operating distance:

$$k_{\min\,typ} = \left(\mu_0 \cdot \frac{r^2}{(r_1{}^2 + d_{\max}{}^2)^{3/2}} \right) \cdot N_1 N_2 s_2 \cdot \left(\frac{1}{\sqrt{L_1 L_2}} \right)$$

When

$k_{\min\,typ} = 0.4\%$

$\mu = \mu_0 = 4 \times \pi \times 10^{-7}$

$N_1 = 1$

$N_2 = 4$

$s_2 = 38.56 \times 10^{-4}\,\mathrm{m}^2$

$L_1 = 0.5\,\mu\mathrm{H}$

$L_2 = 3.6\,\mu\mathrm{H}$

$r_1 = 7.9\,\mathrm{cm}$

we obtain

$d_{\max\,typ} = 21\,\mathrm{cm}, \ldots$ a value which (luckily) matches the one calculated by means of B_{dt}!

Note   Generally, the value of $k_{\min}$ must be at least 0.3%, that is, $k = 0.003$. In this example, a coupling coefficient of $k = 0.3\%$ would have given the following results:

$$V_{20\,min} = V_1 \cdot k_{\min}\sqrt{\frac{L_2}{L_1}}$$

$$= 26.77 \times 0.003\sqrt{\frac{3.6}{0.5}}$$

$$= 215\,\mathrm{mV}$$

and the appearance across the terminals of the integrated circuit of a voltage $V_{ic\,min}$ amounting to

$$V_{ic\,min} = Qp_2 \cdot V_{20\,min}$$

$$= 13.7 \times 0.215$$

$$= 2.95\,\text{V}$$

In this case, the value of the mutual induction would be

$$M_{min} = k\sqrt{\frac{L_2}{L_1}}$$

$$= 0.003\sqrt{\frac{3.6}{0.5}}$$

$$= 4.025\,\text{nH}$$

The maximum possible operating distance would be

$$k_{min} = \left(\mu\frac{r^2}{2(r_1^2 + d_{max}^2)^{3/2}}\right) \cdot N_1 N_2 s_2 \cdot \frac{1}{\sqrt{L_2 L_1}}$$

where

$k_{min} = 0.3\% = 3 \times 10^{-3}$ $\qquad$ $s_2 = 38.56 \times 10^{-4}\,\text{m}^2$

$\mu = \mu_0 = 4 \times \pi \times 10^{-7}$ $\qquad$ $L_1 = 0.5\,\mu\text{H}$

$N_1 = 1$ $\qquad$ $L_2 = 3.6\,\mu\text{H}$

$N_2 = 4$ $\qquad$ $r_1 = 7.9\,\text{cm}$

and therefore

$$d_{max} = 23.4\,\text{cm}$$

IMPORTANT NOTE

The estimation of this distance does not allow for the tolerances of the components and therefore does not yield the guaranteed "minimum" maximum distance. To find this, we need to calculate all the partial derivatives of the above equation. We assume initially that N_1, N_2 have no tolerances ($N_1 = $ one turn and $N_2 = x$ turns will be implemented as a printed circuit!). The same applies to s_2; same reason, same result! So now we just need to examine the tolerances on the values of L_1 and L_2. These tolerances are large, generally of the order of 10% each. The square root of their product is therefore also 10%. In the worst case, the typical minimum maximum distance is 23.4 cm – 10%, that is, approximately 19 cm. Of course, purists will tell you that you must also integrate the tolerances of the current I_1 flowing in the base station antenna, the variations of the voltage supplied to the antenna, the most unfavourable value of V_{ic}, and so on, but in simple terms the most unfavourable case will be of the order of 16 to 17 cm, so that with a Mifare 1 circuit, tuned in this case to 13.56 MHz, we can reliably produce a "proximity" application according to ISO 14 443 – $d_{max} = 16$ cm – by using an 8 cm diameter base station antenna and a k_{min} of 0.3%. For information, if the transponder had been deliberately detuned to 16 MHz, we would have obtained a maximum distance of approximately 10–11 cm.

Intelligent verification   In the case where the coupling coefficient is 0.3%, the following verification method provides another way of understanding the problem.

We know that

$$V_{20} = \frac{M \cdot \omega c}{\sqrt{\left[1 - \left(\frac{\omega c}{\omega s}\right)^2\right]^2 + \left(\frac{L_{2}s \cdot \omega c}{Rp}\right)^2}} \cdot I_1$$

If $\omega c = \omega s$:

$$I_1 = \frac{V_{20}}{M \cdot \omega c} = \frac{\dfrac{V_{ic}}{Qp_2}}{M \cdot \omega c}$$

$$= \frac{0.215}{4.02 \times 10^{-9} \times 85.157 \times 10^{6}}$$

$$= 630 \, \text{mA rms} = \text{OK!}$$

In this case, the magnetic field would be

$$H_0 = \frac{N_1 I_1}{2 \times r_1}$$

$$= \frac{1 \times 0.630}{2 \times 0.079}$$

$$= 4 \, \text{A/m}$$

and

$$B_0 = \mu \cdot H_0$$

$$= 4 \times 3.14 \times 10^{-7} \times 4$$

$$= 5000 \, \text{nT} \quad \text{Q.E.D.}$$

Given that, in our example, B_{dt} is equal to 222 nT, let us calculate B_0/B_{dt}:

$$\frac{B_0}{B_{dt}} = \frac{5000}{222}$$

$$= 22.52$$

We also know that

$$a = \frac{d}{r_1}$$

and therefore

$$\frac{B_0}{B_{dt}} = (1 + a^2)^{3/2}$$

therefore

$$22.52^2 = (1 + a^2)^3$$

and therefore

$$(1 + a^2) = \sqrt[3]{507.26}$$
$$= 7.96$$

therefore

$$a^2 = 6.96$$
$$a = 2.638$$

and therefore

$$d = a \times r_1$$
$$= 2.638 \times 7.9$$
$$= 20.84 \, \text{cm} \quad \text{Q.E.D.}$$

6.2.4 Matters of flux...

Now let us consider the magnetic quantities in this example.

Magnetic flux produced by the base station   We have seen that the "total magnetic flux" Φ_1 produced by the current I_1 flowing in the base station antenna has the value

$$\Phi_1 = L_1 s \cdot I_1$$
$$= 500 \times 10^{-9} \times 0.630$$
$$= 315 \, \text{nWb/m}^2$$

Mutual inductance   The useful flux Φ_2 received at a distance d by the transponder antenna circuit is equal to

$$\Phi_2 = \Phi \text{ from 1 towards } 2 = M \cdot I_1$$

The last value is also equal to the product of the local induction at the transponder and the collecting surface, that is,

$$\Phi_2 = B_{\text{at the tag}} \cdot (N_2 s_2)$$
$$= B_d \cdot (N_2 s_2)$$

By combining the two equations, we can calculate the mutual inductance

$$M_d = \frac{B_d \cdot (N_2 s_2)}{I_1}$$

For example, consider a contactless system whose specified operating distance is, $d = 2r_1 = 15.8 \, \text{cm}$, in other words, for a magnetic induction $B_{15.8} = B_0/11.8$, where

$$B_0 = 5000 \, \text{nT}$$
$$B_{15.8} = \frac{B_0}{11.8}$$
$$= 424 \, \text{nT}$$

$$M_{15.8} = B_{15.8} \frac{N_2 s_2}{I_1}$$

$$= 424 \times 10^{-9} \frac{4 \times 38.56 \times 10^{-4}}{0.630}$$

$$= 10.4 \, \text{nH}$$

Flux received by the transponder   We can now calculate the value of the "useful magnetic flux" captured by the transponder at 15.8 cm:

$$\Phi_2 = B_{15.8} \cdot (N_2 s_2) = \Phi \text{ from 1 towards } 2 = M \cdot I_1$$

$$= 424 \times 10^{-9} \times (4 \times 38.56 \times 10^{-4}) = 10.4 \times 10^{-9} \times 0.63$$

$$= 6.55 \, \text{nWb/m}^2 \, \text{rms}$$

Voltage present at the input of the integrated circuit   Using the formulae given in Chapter 2, we can easily calculate the voltage induced in "no-load" conditions in the transponder winding which has the following value:

$$V_{20} = M \cdot I_1 \cdot \omega$$

$$= (\Phi \text{ from 1 towards } 2) \cdot \omega$$

$$= 10.4 \times 10^{-9} \times 0.630 \times 85.15 \times 10^6$$

$$= 0.557 \, \text{V}$$

giving us the voltage of V_{ic} applied across the input terminals of the integrated circuit:

$$V_{\text{ic}} = Qp_2 \cdot V_{20}$$

$$= 13.7 \times 0.557$$

$$= 7.63 \, \text{V}$$

Note that this important value is the theoretical voltage which would be developed across the terminals of the transponder integrated circuit if it had no internal regulation circuit (see the end of Chapter 6), or if the latter were not operating.

Coupling coefficient of the system   On the other hand, in this particular case, we can write

$$M_{15.8} = k_{15.8}\sqrt{L_1 L_2}$$

therefore

$$k_{15.8} = \frac{M_{15.8}}{\sqrt{L_1 L_2}}$$

$$= \frac{10.4 \times 10^{-9}}{\sqrt{0.5 \times 10^{-9} \times 3.610 \times 10^{-6}}}$$

$$= 0.00775$$

$$= 0.775\%$$

Note that it must be possible to measure this value by the method described in Chapter 10.

Verification:

Using the formula giving the relationship between the useful flux and the total flux (see Chapter 3), let us quickly verify our example; thus:

$$\frac{\Phi_{\text{useful}}}{\Phi_{\text{total}}} = k\sqrt{\frac{L_2}{L_1}}$$

by separately examining the two sides of the equation, we find

$$\frac{\Phi_{\text{useful}}}{\Phi_{\text{total}}} = k\sqrt{\frac{L_2}{L_1}}$$

$$\frac{6.55 \times 10^{-9}}{315 \times 10^{-9}} = 0.775 \times 10^{-2}\sqrt{\frac{3.6 \times 10^{-6}}{0.5 \times 10^{-6}}}$$

$$0.0207 = 0.775 \times 2.683$$

$$0.0207 = 0.0207 \quad \text{Q.E.D.}$$

Figure 6.9a summarizes the transponder system and *Figure 6.9b* summarizes the base station-transponder system.

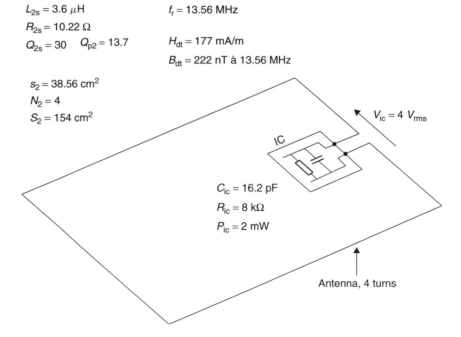

$L_{2s} = 3.6\ \mu\text{H}$ $f_r = 13.56$ MHz

$R_{2s} = 10.22\ \Omega$

$Q_{2s} = 30$ $Q_{p2} = 13.7$ $H_{dt} = 177$ mA/m

$B_{dt} = 222$ nT à 13.56 MHz

$s_2 = 38.56$ cm^2

$N_2 = 4$

$S_2 = 154$ cm^2 $V_{ic} = 4\ V_{rms}$

IC

$C_{ic} = 16.2$ pF

$R_{ic} = 8$ kΩ

$P_{ic} = 2$ mW

Antenna, 4 turns

Figure 6.9a   Summary of the transponder parameters

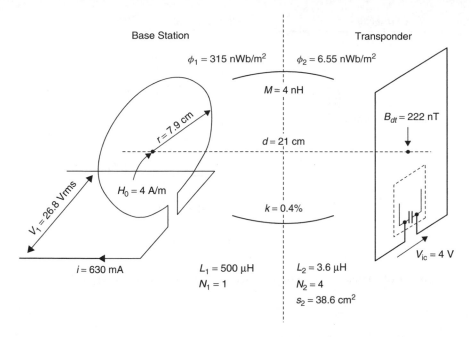

Figure 6.9b Summary of the "base station + associated transponder" system

6.2.5 Return voltage induced in the base station antenna

To conclude our discussion of this application, let us provide an order of magnitude of the voltage induced in the base station antenna during the stage of load modulation of the transponder used to provide the downlink communication from the transponder to the base station.

As mentioned in Chapter 3, by reciprocity, this voltage is

$$\Delta V_1 = \Delta V_2 \cdot k \sqrt{\frac{L_1}{L_2}} \cdot Q_1$$

where ΔV_2 represents the voltage variation (peak to peak) produced by the electronic circuit of the transponder load modulator across the terminals of the antenna winding (see *Figure 6.10*). This variation is usually of the order of 3 to 5 V peak to peak (shown as V pp) in commercially available circuits.

Knowing this value, in the most difficult case – in other words, when the coupling coefficient is lowest – we can put a number on the variation of voltage ΔV_1 developed across the terminals of the base station antenna, which must be extracted from the noise, amplified, filtered, demodulated and decoded. Let us return to the values used before, with

$L_1 = 0.5\,\mu\text{H}$

$L_2 = 3.6\,\mu\text{H}$

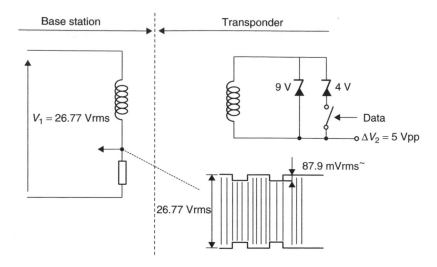

Figure 6.10   Principle of transponder load modulation and its effect on the carrier

$Q_1 = 35$

$k = 0.35 \times 10^{-2}$ (i.e. at $d = 13\,\text{cm}$)

and assume that

$\Delta V_2 = 5\,\text{V pp}$ (square signal)

We find

$$\Delta V_{1\,\text{pp}} = \Delta V_{2\,\text{pp}} \cdot k \sqrt{\frac{L_1}{L_2}} \cdot Q_1$$

$$= 5 \times 0.35 \times 10^{-2} \sqrt{\frac{0.5}{3.6}} \times 35$$

$$= 195.3\,\text{mV pp (square signal)}$$

If we assume that, when received at the base station, the signal is filtered and only its first harmonic is left, we can estimate the effective value of the received signal, namely,

$$\Delta V_{1\,\text{rms}} = \frac{\dfrac{\Delta V_{1\,\text{pp}}}{2}}{\sqrt{2}} \cdot \frac{4}{\pi}$$

$$= \frac{\dfrac{195.3 \times 10^{-3}}{2}}{1.414} \cdot \frac{4}{\pi}$$

$$= 87.9\,\text{mV rms}$$

Assuming that during the response phase, as before, an effective voltage of 26.77 V is present at the base station antenna, we can calculate the voltage ratio "return signal/transmitted signal":

$$\frac{\text{return signal}}{\text{transmitted signal}} = \frac{87.9 \times 10^{-3}}{26.77} = 3.52 \times 10^{-3} = 0.35\%$$

that is, in dB:

$$20 \log(3.52 \times 10^{-3}) = -49.08 \, \text{dB}$$

6.3 "Vicinity" Applications (Approximately 70 cm) of the ISO 15 693 or 18 000-3 Type and Long-range Applications for Vicinity Cards or Item Management

Contactless electronic labels are becoming widely available on the market, and are used increasingly for identification, marking and traceability.

In this new part of the chapter, we will consider some examples of applications relating to this form of vicinity and long-range contactless radio frequency identification (RFID).

6.3.1 The I·CODE integrated circuit

To illustrate our examples operating at 13.56 MHz, we have decided to use Philips Semiconductors I·CODE integrated circuits, as they are among the most commonly used components for these applications. Moreover, this class of integrated circuit (SLI) conforms to the ISO 15 693 standard covering vicinity cards. We should also note that the protocol described in ISO 15 693 currently plays an important part in projects for which standards are being drawn up for item management applications using electronic labels (standard class ISO 18 000-3 dedicated to applications operating at 13.56 MHz).

Table 6.4 briefly summarizes its principal characteristics.

Table 6.4 Characteristics of the I·CODE integrated circuit

Reading/Writing	Write distance equal to read distance
Multitags facility	Anticollision
Excitation frequency	13.56 MHz
Power consumption	200 μW
E2PROM capacity	512 bits
Operating distance	1.2 m multitags L/E (50 × 50 mm label)
	1.5 m detection (EAS)
Transaction speed	20 tags/s
According to current regulations (ETSI 300 330, FCC 47 part 15)	

This completes this short introduction which was simply intended to set out the main reasons for choosing this type of circuit for use in the following examples.

6.3.2   Choosing the quality factors of a vicinity contactless system

Before describing these different examples of applications, let us start by seeing how to approach the choice of quality factors for the base station and transponder. This is because we must make some modifications according to the different applications known as "long range"; we can group these roughly according to the following principles:

— do we want to read/write to transponders at a long distance *only*?

— do we want to read/write to transponders, *which can be individually located at a short distance as well*?

— do we want, in general, to read/write to transponders *located simultaneously at short and long distances*?

What is this for?
As will be shown in the following sections, the answers to these questions are directly related to the physical relationships between the upward and downward induced voltages, namely,

— for the downlink, from the base station to the transponder:

$$V_2 = V_1 \cdot M \cdot \omega \cdot I_1$$

$$= V_1 \cdot k \sqrt{\frac{L_2}{L_1}}$$

$$= Qp_2 \cdot V_{20}$$

— for the uplink, from the transponder to the base station, the inverse formula:

$$\Delta V_1 = Q_1 \left(\Delta V_2 \cdot k \sqrt{\frac{L_1}{L_2}} \right)$$

In the transponder   As mentioned before, where the uplink is concerned, in order to avoid large variations in the magnetic field threshold voltage, the quality factor for the transponder under load, Qp_2, must be as high as possible (see Chapter 2 again if necessary), and we must then try to optimize its value for optimal energy transfer.

In the base station   For the uplink, from the base station to the transponder, the situation is quite different.

Everything else being equal, an increase in the quality factor of the base station antenna Q_1 is accompanied by an increase in the current flowing in the base station antenna, and therefore in the radiated magnetic field, thus providing better remote supply and operation of the transponder over a longer range. We should therefore use a value of Q_1 which is as high as possible. We have seen (Chapter 3) that the maximum value chosen by the industry is of the order of a hundred. Everything else being equal, this is summarized by the equation

$$V_2 = V_1 \cdot M \cdot \omega \cdot I_1 = Q_1 \cdot \ldots$$

In fact, this maximum value of Q_1 is only applicable to systems communicating with transponders which are all located at long distances, since in this case the value of V_2 must be maximized for the successful remote supply of the integrated circuit.

Now we shall look at the downlink, from the transponder to the base station.

The formula below shows the parameters used for calculating the voltage developed in the base station, following the load modulation of the transponder:

$$\Delta V_1 = Q_1 \cdot \Delta V_2 \cdot k \sqrt{\frac{L_1}{L_2}}$$

We need to consider two cases.

— The first is that of systems designed to communicate with transponders located at long distances only. In this case, for a given coupling factor k, we must maximize all the parameters and therefore maintain a high value of Q_1, as described in the previous sections, of the order of a hundred.

— The second case is that of systems designed to communicate with and process information from transponders located at both short and long distances. The latter situation gives rise to problems related to the sensitivity of the demodulator and the width of the bandwidth of the base station antenna circuit.

Sensitivity of the demodulator   The input stage of the amplifier preceding the demodulation stage of the base station must have a high sensitivity to operate at long distance, but must not become saturated in short-distance operation. To resolve this double problem, use is generally made of an amplification device using automatic gain control (AGC), which, unfortunately, does not have an infinite operating range. To help to overcome this problem relating to the AGC, a compromise is commonly reached, by reducing the value of Q_1. This is generally located in the area of $Q_1 = 35$.

Bandwidth of the base station antenna circuit   As the value of Q_1 rises, the bandwidth of the base station antenna circuit decreases ($\Delta F = f_0/Q_1$). Unfortunately, during the active phase of the downlink, the transponder modulates its load in an attempt to ensure that it is understood by the base station. This modulation, as mentioned above, creates sidebands, containing the return information energy, in the spectrum of the signal sent towards the demodulator. If these signals are to be recovered, the tuned antenna circuit must not be too selective. This becomes even more important when the original signals are weak, and therefore mainly in the case of long-distance applications (see *Figure 6.11*). Generally, a value of $Q_1 = 35$ is equally suitable ... but this is only a happy coincidence because the choice of the values to be used for Q_1 – in

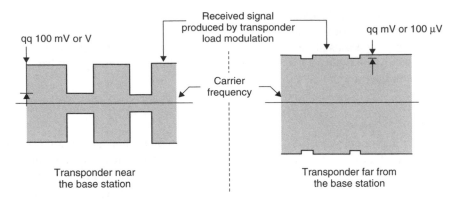

Figure 6.11   Modulation created by the transponder near and far from the base station

the base station – and Qp_2 – in the transponder – is a clever compromise which must allow for all these parameters. There is no universal answer to these problems, and the global transfer function of each application must be examined separately for each case if satisfactory results are to be achieved.

6.3.3   First example – base station, for "long-range" applications only

The first example described below relates to an application operating at long range only, and therefore $Q_1 = 100$.

In this example, we shall start from the working hypothesis that the size of the antenna is specified in advance, to limit the overall dimensions. Let us consider, for example, a circular antenna with a diameter of 60 cm.

Mechanical characteristics of the base station antenna   The mechanical characteristics of the base station antenna are shown in *Table 6.5*.

Radiation characteristics of the base station antenna   Radiation resistance of the antenna:

$$Ra = \frac{31\,200}{\lambda c^4} S_1{}^2$$

$$= \text{const} \cdot S_1{}^2$$

$$= 0.13 \times 798.6 \times 10^{-4}$$

$$= 10.30\,\text{m}\Omega$$

Electrical parameters of the physical circuit of the base station antenna   We shall proceed in the same way as for the short-range (proximity) example, using the MIFARE circuit.

Inductance, resistance and capacitance of the antenna   Let us move directly to the measured values.

Table 6.5

Number of turns	N_1	$= 1$
Geometry of the circular turn(s)	Diameter $D_1 = 60\,\text{cm}$	
	radius r_1	$= 30\,\text{cm}$

If the antenna is square $(a \times a)$ or rectangular $(a \times b)$, then, as a first approximation, we can calculate the radius as equivalent to that of a circular antenna $r_{\text{equi.}} = \sqrt{\dfrac{a \cdot b}{\pi}}$.

Length of one turn	l_1	$= 2 \times \pi \times r_1$
		$= 2 \times 3.14 \times 0.3$
		$= 1.884\,\text{m}$
Surface area of one turn of the antenna	s_1	$= \pi \times \dfrac{D_1{}^2}{4}$
		$= 3.14 \times \dfrac{0.6^2}{4}$
		$= 2826 \times 10^{-4}\,\text{m}^2$
Total surface area of the antenna	S_1	$= N_1 s_1$
		$= 1 \times s_1$
	$S_1{}^2$	$= (N_1 s_1)^2$
		$= 798.6 \times 10^{-4}\,(\text{m}^2)^2$

Examples of measured values

inductance $L_1 s = 1.8\,\mu\text{H}$

series resistance $R_1 s = 0.9\,\Omega$

parallel capacitance $C_1 p = 8.25\,\text{pF}$

The equivalent circuit of the antenna is shown in *Figure 6.12*.

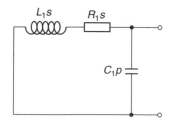

Figure 6.12 Series equivalent circuit of the base station antenna

Intrinsic resonant frequency of the antenna, fs Assuming that $\omega s = 2\pi fs$, we can calculate the intrinsic frequency of the base station antenna. *fs* is such that

$$L_1 s \cdot C_1 p \cdot \omega^2 = 1$$

therefore

$$\omega s^2 = \frac{1}{14.85} 10^{18}$$

$$\omega s = 259 \times 10^6$$

therefore

$$fs = 41.24\,\text{MHz}$$

Impedance of the base station antenna inductance   At 13.56 MHz, the inductance of the base station antenna has an impedance of:

$$Z_{L_1 s} = L_1 s \cdot \omega c$$

$$= 1800 \times 10^{-9} \times 85.157 \times 10^6$$

$$= 153.28\,\Omega$$

Intrinsic quality factor of the base station antenna   In the same way as before,

$$Q_{L_1 s} = \frac{L_1 s \cdot \omega c}{R_1 s} = \frac{153.28}{0.9} = 170.31$$

Application-specific parameters   To meet the requirements of the proposed application, in terms of its bit rate and its temporal characteristics (e.g. the pause time in 100% ASK modulation), the value of the quality factor of the system must be adapted (i.e. reduced); thus, in the case of the I·code class, we must operate with a value of $Q_{1\,\text{max}}$ of approximately 100 under load for applications operating only, as mentioned above, over long distances.

To decrease the value of $Q_1 s$ to the value of Q_1 under load at 13.56 MHz, we must again connect the antenna in series with an additional resistance R_{ext}. As a first approach, assuming that the value of $C_1 p$ is negligible, we can write

$$Q_{1\,\text{load}} = 100 \Rightarrow \text{long-range applications only denoted as } Q_1 \text{ subsequently.}$$

$$Q_1 = \frac{L_1 s \cdot \omega}{R_1 s + R_{\text{ext}}}$$

and therefore

$$R_{\text{ext}} = \frac{L_1 s \cdot \omega}{Q_1} - R_1 s$$

$$= \frac{153.28}{100} - 0.5$$

$$= 1.03\,\Omega$$

Parallel equivalent circuit of the physical antenna of the base station   Using a conventional series/parallel transformation (see *Figure 6.13*), we obtain the parallel equivalent

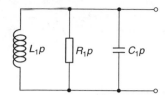

Figure 6.13   Parallel equivalent circuit of the base station antenna

circuit of this whole system, in the form of a conventional anti-resonant circuit:

$$L_1 p = L_1 s = 1800\,\text{nH}$$

$$R_1 p = Q_1^2 \cdot (R_1 s + R_{\text{ext}})$$

$$= 100^2 \times (0.5 + 1.03)$$

$$= 15\,300\,\Omega$$

$$C_1 p = 8.25\,\text{pF}$$

Matching the source impedance (amplifier) and load impedance (base station antenna circuit) and EMC problems   The global impedance of the anti-resonant circuit forming the inductance of the base station antenna and its tuning capacitance is not adapted in any way (in terms of power) to the output impedance (real only) of the power amplifier that drives the antenna, whose value is $R_{\text{out}} = 50\,\Omega$.

As before, we have to provide this impedance matching. This is done with the aid of an electrical circuit based on a system forming a capacitance bridge.

Additionally, in the case of long-range applications (50 to 70 cm), which by definition require higher radiated powers than those encountered in proximity systems, better adapted systems, in other words structurally less radiating systems, must be used in order to overcome the problems of electromagnetic pollution (EMC) and comply with the associated standards (ETSI 300 330 and/or FCC 47, part 15).

The trend is to use a balun system designed to balance the electrical radiation with respect to earth, while making the electrical signals applied to the base station antenna symmetrical (see the theoretical information in Chapter 3).

Thus, the global circuit including the "balun + capacitance bridge" system takes on a different configuration, as shown in *Figures 6.14a* and *6.14b*.

The physical capacitance to be added to the circuit, C_2, must also be incorporated in the new value C_p' of C_p, and the whole load formed in this way must be expressed with respect to the transformer primary to estimate the load which it represents for the amplifier ($C_p' = C_p + C_2$), giving, for the values x' with respect to the primary

$$R' = \frac{Rp}{n^2} = \frac{15.3 \times 10^3}{4} = 3.825\,\text{k}\Omega$$

$$L_1'p = \frac{Lp}{n^2} = \frac{1.8 \times 10^{-3}}{4} = 450\,\text{nH}$$

$$C' = (C_p + C_2) \cdot n^2 = (8.25 \times 10^{-12} \times 4) + (C_2 \cdot n^2)$$

$$= 33\,\text{pF} + (C_2 \cdot n^2)\,\text{pF}$$

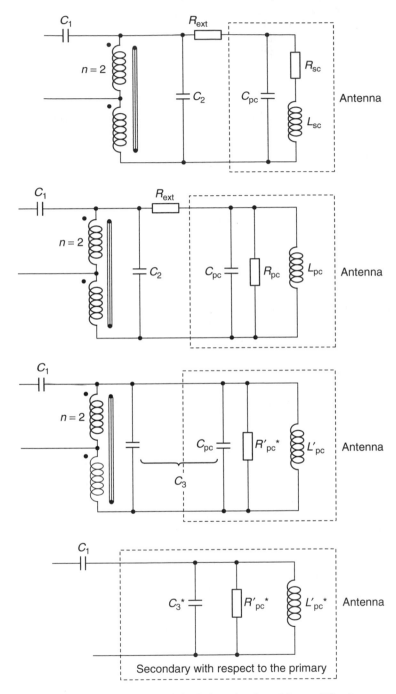

Figure 6.14a Diagram of the balun circuit and its modifications

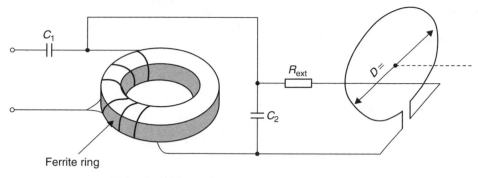

Figure 6.14b Formation of a balun transformer on a ferrite ring

To meet these matching conditions, the values of the two capacitances must then be

$$C_1 = \frac{1}{\omega \cdot \sqrt{R_{\text{out}} \cdot R'}}$$

$$\frac{1}{L_1' p \cdot \omega^2} = C_1 + C' = C_1 + (Cp + C_2)n^2$$

$$n^2 C_2 = \frac{1}{L_1' p \cdot \omega^2} - C_1 - Cpn^2$$

therefore

$$C_1 = \frac{1}{85.156 \times 10^6 \times \sqrt{50 \times 3825}}$$
$$= 26.7 \, \text{pF}$$

$$4C_2 = \frac{1}{(450 \times 10^{-9}) \cdot (85.156 \times 10^6)^2} - 26.7 - (8.25 \times 4)$$

and therefore

$$C_2 = 54 \, \text{pF}$$

Electrical stresses on each element All the values of the different elements are then known. We must now estimate the electrical stresses (current/voltage/power), which all these elements have to withstand.

NOTE
The power in watts is retained.

The electrical stresses Assuming that the amplifier delivers a known maximum power, for example,

power delivered by the amplifier $P_{1 \, \text{max}} = 4 \, \text{W}$ rms, then

— input impedance (real only):

$$R_{\text{in}} = R_{\text{out}} = 50\,\Omega$$

— input current:

$$I_{\text{in rms}} = \sqrt{\frac{P_1}{Z_{\text{in}}}} = 0.282\,\text{A rms}$$

$$I_{\text{in}}p = 0.282\sqrt{2} \approx 400\,\text{mA}$$

Capacitance:

— current flowing through C_1:

$$I_{C_1} = I_{\text{in}}$$

$$= 0.282\,\text{A rms}$$

— voltage across the terminals of C_1:

$$U_{C_1} = \frac{1}{C_1 \cdot \omega} I_{C_1}$$

$$= \frac{1}{43 \times 10^{-12} \times 85.157 \times 10^6} 0.109$$

$$= 27.3\,\text{V rms}$$

Antenna:

— Current in the antenna:

$$I_{L_1 s} = \sqrt{(P_1 \cdot Q_1) \cdot (L_1 s \cdot \omega)}$$

$$= \sqrt{(4 \times 100) \times (1.800 \times 10^{-6} \times 85.157 \times 10^6)}$$

$$= 1.615\,\text{A rms}$$

— Voltage across the terminals of L (antenna):

$$U_{L_1 s} = (L_1 s \cdot \omega) \cdot I_{L_1 s}$$

$$= 155.28 \times 1.615$$

$$= 247.6\,\text{V rms}$$

$$U_{L_1 s\ \text{pp}} = 2 \times 247.6 \times 1.414$$

$$= 700.16\,\text{V pp}$$

R_{ext}:

—   Current flowing in R_{ext}

$$I_{R_{\text{ext}}} = I_{L_1 s}$$

—   Power dissipated in R_{ext}

$$P_{R_{\text{ext}}} = R_{\text{ext}} \cdot I_{L_1 s}^2$$

Values of magnetic fields, induction and flux, and radiated power

Magnetic field in the centre of the antenna   Value of the magnetic field created in the centre of the antenna:

$$H_0 = \frac{N \cdot I_{L_1 s}}{2r_1}$$

$$= \frac{1 \times 1.615}{2 \times 0.3}$$

$$= 2.69 \, \text{A/m}$$

Magnetic induction in the centre of the antenna   The associated value of the magnetic induction is

$$B_0 = \mu_0 \cdot H_0$$

$$= 12.56 \times 10^{-7} \times 2.69$$

$$= 33.79 \, \mu\text{T rms}$$

For information, I will now assign some orders of magnitude to the field and induction at 60 cm:

—   the magnetic field produced in the air at $d = 2r_1 = 60 \, \text{cm}$ is

$$H_{60} = \frac{H_0}{11.8}$$

$$= \frac{2.69}{11.8}$$

$$= 227 \, \text{mA/m}$$

—   the induction produced in the air at $d = 2r_1 = 60 \, \text{cm}$:

$$B_{60} = \frac{\mu_0 \cdot H_0}{11.8}$$

$$= \frac{12.56 \times 10^{-7} \times 2.69}{11.8}$$

$$= 286.3 \, \text{nT}$$

Radiated power

—   power radiated by the antenna:

$$Pa = Ra \cdot I_{L_1s}{}^2$$

$$= 10.38 \times 10^{-3} \times 1.615^2$$

$$= 27\,\text{mW}$$

Table 6.6 summarizes the values of the different parameters.

Brief verification and comment   If the power in the load is 4 W, this power P_1 is dissipated only in the resistance Rp connected in parallel with the tuned circuit of the base station, and therefore, because $P_1 = R_1p \cdot Ip^2$,

$$Ip = \sqrt{\frac{P_1}{R_1p}}$$

$$= \sqrt{\frac{4}{15\,300}}$$

$$= 16.16\,\text{mA rms}$$

The current flowing in the base station antenna coil is therefore as follows:

on the one hand $I_{\text{ant}} = Q_1 \cdot Ip$

$$= 100 \times 16.16 \times 10^{-13}$$

$$= 1.616\,\text{A rms}$$

and on the other hand $I_{\text{ant}} = \sqrt{\dfrac{P_1 \cdot Q_1}{L_1s \cdot \omega}}$

$$= \sqrt{\frac{4 \times 100}{153.29}}$$

$$= 1.616\,\text{A}\quad\text{Q.E.D.}$$

Table 6.6

Base station antenna	Balun	
$L_1s = 1.8\,\mu\text{H}$	P_1	$= 4\,\text{W}$
$R_1s = 0.9\,\Omega$	I_1	$= 1.615\,\text{A}$
$N_1 = 1$	V_1	$= 247.6\,V\,\text{rms}$
$D_1 = 60\,\text{cm}$	H_0	$= 2.69\,\text{A/m}$
$r_1 = 30\,\text{cm}$	B_0	$= 33.79\,\mu\text{T}$
$s_1 = 2826\,\text{cm}^2$	H_{60}	$= 227\,\text{mA/m}$
$Q_1 = 100$ under load	Pa	$= 27\,\text{mW}$
$Ra = 10.3\,\text{m}\Omega$		

In fact, the power delivered by the amplifier is slightly greater, since the resistance R_{ext} connected in series with the antenna circuit also consumes some of the power:

$$P_{R_{ext}} = R_{ext} \cdot I^2$$
$$= 1.03 \times 1.616^2$$
$$= 2.69\,\text{W}$$

therefore

$$P_{total} = 4 + 2.69$$
$$= 6.69\,\text{W}$$

VERY IMPORTANT NOTE

According to the last line of *Table 6.6*, the radiated power is 27 mW. And why not? you may ask. And you will be right! In fact, European (CEPT-ERC 70 30) and French (ART publications) regulations specify no constraints in respect of maximum emitted power for base stations for RFID applications operating at 13.56 MHz... but (there is always a "but", and a considerable one in this case) we must also comply with ETSI 300 330. What does this mean?

The constraints of ETSI 300 330   European law, imposes a number of constraints on the use of short-range devices (SRD) known as "non-specific", such as those used in RFID applications in the 13.56 MHz ISM band. (The values are closed to FCC.)

Of all these constraints, what matters to us at this point is pollution prevention in the radio frequency spectrum. Anyone wanting more details of all these subtleties should again consult my earlier book. Returning to our initial theme, in the context of operation at 13.56 MHz, the essential standard to be observed is the one issued by ETSI under reference number 300 330 USA FCC 47, part 15. This stipulates that, in specific conditions of measurement (measurement methods, conformity with limits, etc.), the magnetic field radiated at 10 m by the base station antenna, regardless of the radiated power, must not exceed $H_{\text{max ETSI 10 m}} = 42\,\text{dB}\mu\text{A/m}$. After the usual conversions (see the book cited above), this (remote) magnetic field value can be converted to an equivalent electrical field having the following value:

$$E_{\text{max ETSI 10 m}} = [H_{\text{max ETSI 10 m}}(\text{en dB}\mu\text{A/m}) + 51.5]\ (\text{in dB}\mu\text{V/m})$$
$$= (42 + 51.5)\,\text{dB}\mu\text{V/m}$$
$$= 93.5\,\text{dB}\mu\text{V/m}$$

and therefore

$$E_{\text{max ETSI 10 m}} = 47.4\,\text{mV/m}$$

Additionally, for conventional antennae (circular, rectangular, flat, etc.) it is demonstrated that the remote electrical field is proportional to the square root of the radiated power and inversely proportional to the distance, and is essentially expressed by the following equation:

$$E \approx \frac{7\sqrt{Pa}}{d}$$

This equation can be used to calculate the maximum value of Pa_{max} representing the "theoretically official" boundary of the well-known "42 dBμA/m at 10 m" of the ETSI standard, that is,

$$Pa_{max} \approx \left(\frac{E_{max\,ETSI} \cdot d}{7} \right)^2$$

$$\approx \left(\frac{47.4 \times 10^{-3} \times 10}{7} \right)^2$$

$$\approx 4.585\,mW$$

So that's settled.

No doubt you noticed, a long way back, that we established that the value of Pa_{max} used in the systems described previously was often about 10 mW, and not 4.585 mW as indicated above. Leaving aside the fact that the calculations are easier to understand and deal with if we use 10 mW instead of 4.585 mW, you should know that there are many clever ways of designing base station antennae (see Chapter 8, on "figure of 8" coil antennae, for example) that enable us, while scrupulously observing the ETSI stipulation of "42 dBμA/m at 10 m", to increase the radiated power Pa to approximately 10 mW, and consequently to increase the operating range of a system which operates in the near field rather than the far field area.

Clearly, if the system exceeds the maximum power authorized by regulation, it will be necessary to contact the Regulatory authority to obtain either a derogation or an authorization with a site licence.

If we wish to operate without a licence, in other words without exceeding the approximate level of $Pa_{max} = 5$ to 10 mW, then, given the same mechanical antenna (and thus the same value of Ra), we must consider decreasing the power delivered by the amplifier, so that, in the original example given above,

$$Pa = Ra \cdot I_{L_1s}^{\,2}$$

$$10 \times 10^{-3} = 10.38 \times 10^{-3} \times I_{L_1s}^{\,2}$$

and therefore

$$I_{L_1s} = \sqrt{\frac{10}{10.38}}$$

$$= 0.98\,A$$

In this case, the magnetic field strength at the centre is

$$H_0 = \frac{N_1 \cdot I_{L_1s}}{2r_1}$$

$$= \frac{1 \times 0.98}{2 \times 0.3}$$

$$= 1.6\,A/m$$

and H_{60} is as follows:

$$H_{60} = H_{2r1} = \frac{H_0}{11.8}$$

$$= \frac{1.6}{11.8}$$

$$= 135\,\text{mA/m}$$

In this case,

$$H_{60} = \frac{H_0}{11.8}$$

$$= \frac{1.6}{11.8}$$

$$= 135.6\,\text{mA/m}$$

On the other hand,

$$I_{L_1 s} = \sqrt{\frac{P_1 \cdot Q_1}{L_1 s \cdot \omega}}$$

and therefore

$$P_{1\ \text{max ampli}} = \frac{L_1 s \cdot \omega \cdot I_{L_1 s}{}^2}{Q_1}$$

$$= \frac{1.8 \times 10^{-6} \times 85.157 \times 10^6 \times 0.98^2}{100}$$

$$= 1.35\,\text{W}$$

This is rather a crude way of resolving the problem. Now let us see how we can optimize the system in the case where the radiated power Pa is specified in advance.

6.3.4 Second example (or another way of looking at the problem)

We shall stay with the preceding example, but now examine the base station from another angle, where our objective is **a known value of radiated power,** $Pa = \text{const.}$

We shall take a specific example, setting the target of not exceeding a maximum power EIRP of $Pa = 10\,\text{mW}$ rms with the following additional assumptions:

$N_1 = 1$

$r_1 = ?$ cm, the value to be defined.

$H_{2r1} = \dfrac{H_0}{11.8} = 100\,\text{mA/m}$ (the conventional value for this kind of label).

and therefore

$$H_0 = 1.18 \times H_{2r1}$$

$$= 1.18\,\text{A/m rms}$$

therefore

$$B_0 = \mu \times H_0$$

$$= 1.5\,\mu\text{T rms}$$

$$B_{2r1} = \frac{1.5\,\mu\text{T}}{11.8}$$

$$= 127\,\text{nT}$$

As shown in the short-range example,

$$I_1{}^6 = \frac{Pa \cdot (2H_0)^4}{\text{const} \times \pi^2 N_1{}^6}$$

$$= \frac{10 \times 10^{-3} \times (2 \times 1.18)^4}{0.13 \times 3.14^2 \times 1^6}$$

$$= 0.242$$

and therefore

$$I_1 = 0.79\,\text{A rms}$$

We can also calculate r_1, thus:

$$r_1 = \frac{N_1 I_1}{2H_0}$$

so that $r_1 = 33.4\,\text{cm} \Rightarrow \text{diameter} = 66.8\,\text{cm}$

Knowing r_1, we can then calculate the field strength H_d and induction B_d created at a distance d, by means of the two equations $d = a \cdot r_1$ and

$$B_d = \frac{B_0}{(1 + a^2)^{3/2}}.$$

Verification:

Surface area of one turn:

$$s_1 = \pi r_1{}^2$$

$$= 3.14 \times (0.334)^2$$

$$= 3845\,\text{cm}^2$$

and therefore

$$S_1 = N_1 s_1$$

$$= 1 \times 3845$$

$$= 3845\,\text{cm}^2$$

therefore

$$Pa = \text{const} \cdot S_1{}^2 \cdot I_{\text{ant}}{}^2$$

$$= 0.13 \times (3845 \times 10^{-4})^2 \times 0.79^2$$

$$= 10\,\text{mW}$$

If we continue to use the same physical antenna as that defined in the description above, where

$$L_1s = 1.8\,\mu\text{H}$$

$$R_1s = 0.9\,\Omega$$

the voltage across the terminals of the base station antenna will be

$$
\begin{aligned}
U_{L_1s} &= Z_{L_1s} \cdot I_{\text{ant}} \\
&= L_{1s} \cdot \omega \cdot I_{\text{ant}} \\
&= (1.8 \times 10^{-6} \times 85.157 \times 10^{6}) \times 0.79 \\
&= 122.6\,\text{V rms or, again, } 346.8\,\text{V pp}
\end{aligned}
$$

and the magnetic flux produced by the base station antenna will be

$$
\begin{aligned}
\Phi_1 &= L_1s \cdot I_{\text{ant}} \\
&= 1.8 \times 10^{-6} \times 0.79 \\
&= 1440\,\text{nWb/m}^2
\end{aligned}
$$

6.3.5 Third example (or yet another way of looking at the same problem)

Let us evaluate the power in watts – P – required to obtain a given power Pa. Specifying, for example, that

$$Pa = 10\,\text{mW}$$

and

const	$= 0.13$
Q	$= 100$
H_0	$= 1.18\,\text{A/m (a value below that of the ISO 15\,693 standard)}$
L_1s	$= 1.8\,\mu\text{H}$
$L_1s \cdot \omega$	$= 153.28\,\Omega$
N_1	$= 1$

We demonstrated previously that

$$P_1{}^3 = \frac{Pa(2H_0)^4 \cdot (L_1s \cdot \omega)^3}{\text{const} \cdot N_1{}^6 \pi^2 Q_1{}^3}$$

therefore

$$P_1{}^3 = \frac{10 \times 10^{-3} \times (2 \times 1.18)^4 \times 153.28^3}{0.13 \times 1^6 \times 3.14^2 \times 100^3}$$

$$= 0.875$$

and therefore

$$P_1 = 0.954 \, \text{W rms}$$

A quick check:

$$I_1 = \sqrt{\frac{P_1 \cdot Q_1}{L_1 s \cdot \omega}}$$

$$= \sqrt{\frac{0.954 \times 100}{153.28}}$$

$$= 0.954 \, \text{A} \qquad \text{Q.E.D.}$$

NOTE

For a system operating at short AND long range, and therefore using a value of $Q_1 = 35$ (in place of the 100 used before), we would find

$$P_1{}^3 = 20.4$$

and therefore

$$P_1 = 2.71 \, \text{W rms}$$

and

$$I_1 = \sqrt{\frac{2.71 \times 35}{153.28}}$$

$$= 0.79 \, \text{A}$$

$= \ldots$ the same as before because there was no reason for the field strength H_0 to change!

If the amplifier also operates on a matched load of 50 Ω, we can estimate the voltage that must be supplied to it. This is because

$$P = \frac{U^2}{R}$$

$$1 = \frac{U^2}{50}$$

and therefore

$$U = 7.1 \, \text{V rms}$$

This sinusoidal voltage is what has to be applied to the input of the system.

If the amplifier delivers square signals, we know (Chapter 3) that

$$V_{square} = \frac{\pi}{4} V_{sine}$$

$$= 0.785 \times 7.1$$

$$= 5.6\,V$$

This means that the output stage of the amplifier could be a bridge circuit supplied at approximately 7 to 8 V (5.6 V plus the breakdown voltages of the various transistors)... or a class C circuit.

We can also estimate the current flowing in the load.

$$P = UI$$

$$1 = 5.6 \times I$$

and therefore

$$I = 0.18\,A$$

Table 6.7 summarizes the principal characteristics, for the base station.

Table 6.7

Optimization of the parameters of the base station, for a maximum radiated power of 10 mW

Pa = 10 mW

Base station antenna

L_1s	= 1.8 µH	
R_1s	= 0.9 Ω	
N_1	= 1	
D_1	= 66.8 cm	
r_1	= 33.4 cm	
s_1	= 0.3503 m²	
Q_1	= 100 Long range	$(Q_1 = 35)$ short AND long range
Ra	= 15.95 mΩ	
P_1	= 0.954 W	$(P_1 = 2.71\,W)$
I_1	= 0.79 A	$(I_1 = 0.79\,A)$
V_1	= 122.6 V rms	
H_0	= 1.18 A/m	$(H_0 = 1.18\,A/m)$
B_0	= 1.5 µT	$(B_0 = 1.5\,µT)$
$H_{66.8}$	= 100 mA/m	

After this summary of the various parameters of the base station, let us move on to the optimization of the transponder parameters.

6.3.6 Seen from the transponder

Regardless of the distance (responsible for the local induction $B(d, r)$ present in the vicinity of the transponder), a voltage V_{ic} (according to the specifications of the integrated circuit) has to be applied to the terminals of the integrated circuit to make it operate correctly.

The principal characteristics of the I·CODE integrated circuit are shown in *Table 6.8* (see also the copy of the data sheet values in *Figure 6.15*).

Table 6.8

Characteristics of the integrated circuit

$C_{ic\,typ}$ $= 23.2\,pF$ at 2.2 V rms
with some typical values by way of supplementary information:
C_{coil} $= 5\,pF$
$C_{parasite}$ $= 5\,pF$
$C_{mounting} = 2\,pF$
therefore
$C_{total\,typ}$ $= 35.2\,pF$
$R_{ic\,typ}$ $= 24.2\,k\Omega$ at 2.2 V rms
$P_{min\,typ}$ $= 200\,\mu W$
$V_{ic\,typ}$ $= 2.2$ V rms
$R_{th\,j\text{-}amb}$ $\approx 50°C/W$
i_{max} $= 30\,mA$ rms max
(maximum value, not to be exceeded in a standard application)
$i_{max\,rating} = \pm 60\,mA$ peak
(maximum value not to be exceeded in case of a fault in a standard application)

NOTES

— The values of C_{ic} and R_{ic} must be measured with a network analyser (Hewlett Packard, for example) for a specified sinusoidal alternating voltage V_{ic}.

— The voltage $V_{ic\,typ}$ is the typical voltage required for the correct operation of the transponder.

— To ensure the satisfactory operation of an application, we must use the value $V_{ic\,max}$ (of the order of 3.5 V rms), which, depending on variations due to the production process, is the voltage that must be applied to the "quietest" integrated circuit to make it start to operate correctly.

— The typical minimum power consumption shown is calculated by means of the formula $P_{ic} = (V_{ic\,typ}^2)/(R_{ic})$. The "minimum power consumption" means the power in watts consumed by the integrated circuit when it starts to be active; in other words, when it consumes this power it becomes active. Since this value is specified as 2.2 V rms, we can conclude that the circuit has an equivalent resistance of approximately

$$R_{ic} = \frac{2.2}{200 \times 10^{-6}} = 24.2\,k\Omega$$

Characteristics of the I·CODE integrated circuit	
Read/Write	Write distance equal to read distance
Multitag option	Anti-collision
Excitation frequency	13.56 MHz
Power consumption	200 μW
EEPROM capacity	512 bits
Operating range	1.2 m multitags R/W (50 × 50 mm label)
	1.5 m detection (EAS)
Transaction rate	20 tags/s
Conforms to current regulations (ETSI 300 330, FCC 47 Part 15)	

Figure 6.15 Characteristics of the Philips SC I·CODE circuit

or that the current flowing in this resistance is as follows:

$$I_{R_{ic}} = \frac{V_{ic}}{R_{ic}} = \frac{2.2}{24.2 \times 10^3} = 90\,\mu\text{A}$$

We should also note that, when the circuit approaches the base station antenna, its power consumption increases, because the internal regulation circuit comes into operation, and in this case we must be careful not to exceed the maximum effective value of the integrated circuit input current (30 mA rms max.) and not to exceed the maximum junction temperature of +85°C (see the description of thermal factors below). You will find a full explanation of these matters at the end of this chapter.

The manufacturer of the integrated circuit also provides some recommendations for use (see *Table 6.9*):

NOTE
In our case, for ease of understanding and to simplify the subsequent calculations, we shall assume that the sum of the connection capacitances, and so on, is 4.4 pF ($C_{tot} = 27.6$ pF), so that, in an application in which the transponder is tuned to the carrier $fr = fc$, we can use a "rounded" value for the inductance of the transponder antenna, namely, $L_2s = 5\,\mu$H. In fact,

$$L_2s(C_{ic} + C)(2\pi \times 13.56)^2 = 1$$

$$5 \times (23.2 + 4.4)(2 \times 3.14 \times 13.56)^2 = 1$$

Table 6.9

Recommendations for use:

Qp_2 at 13.56 MHz > 60 see Chapter 2 on the threshold field strength

If the transponder is tuned to the carrier frequency (13.56 MHz):

$$L_2 s_{\text{typ max}} \cdot C_{\text{ic typ}} \cdot \omega^2 = 1$$

$$\Rightarrow L_2 s_{\text{typ max}} = \frac{1}{23.2 \times 10^{-12} \times 7251.55 \times 10^{12}} = 5.95\,\mu\text{H}$$

$$L_2 s_{\text{typ min}} \cdot C_{\text{total typ max}} \cdot \omega^2 = 1$$

$$\Rightarrow L_2 s_{\text{typ min}} = \frac{1}{35.2 \times 10^{-12} \times 7251.55 \times 10^{12}} = 3.91\,\mu\text{H}$$

therefore $3.9\,\mu\text{H} < L_2 s_{\text{typ}} < 5.9\,\mu\text{H}$ or, again, a mean value of $L_2 s = 4.9\,\mu\text{H}$

Example

To summarize, for a transponder fitted with an I·CODE circuit for operation at $fp = 13.56\,\text{MHz}$, we have adopted the following:

$$\omega c = 2 \times 3.14 \times 13.56 \times 10^6 = 85.157 \times 10^6 \,\text{rd/s}$$

$$L_2 s = 5\,\mu\text{H}$$

$$C_{\text{tot}} = 23.2 + 4.4 \qquad\qquad = 27.6\,\text{pF}$$

In this case, the transponder is tuned to the carrier frequency of the base station

— the integrated circuit alone:

$$V_{\text{ic min}} = 2.2\,\text{V rms}$$

$$R_{\text{ic}} = 24.2\,\text{k}\Omega \text{ at } 2.2\,\text{V}$$

The "energy generator/load" pair   The "transponder antenna circuit/integrated circuit" system forms, as shown previously, an "energy generator/load" pair. As emphasized several times before, when $fr = fc$, the optimal energy transfer takes place when

$$Zi = R_{\text{ic}} = L_2 s \cdot \omega c \cdot Q_2 s$$

or in other words, in the ideal case,

$$Q_2 s = \frac{R_{\text{ic}}}{\omega c \cdot L_2 s}$$

In the example in question, the value of R_{ic} is 24.2 kΩ, and therefore, in order for Zi to be equal to R_{ic}, the quality factor of the antenna alone, $Q_2 s$, must be

$$Q_2 s = \frac{24.2 \times 10^3}{85.157 \times 10^6 \times 5 \times 10^{-6}}$$

$$= 56.85$$

On the other hand, we know (see Chapter 2 again if necessary) that, in order to obtain the maximum value $Q_{2s\max}$ of the quality factor for the global load, we must satisfy the relation

$$L_{2s} \text{ (to find } Q_{2s\max}) = \frac{\sqrt{R_{2s} \cdot R_{\text{ic}}}}{\omega c}$$

Since we have specified the value of L_{2s} and we know R_{ic}, we can use the above equation to calculate the required value of R_{2s} to provide, finally, the maximum value of Q_{p2}, that is,

$$R_{2s} = \frac{(L_{2s} \cdot \omega c)^2}{R_{\text{ic}}}$$

$$= \frac{(5 \times 10^{-6} \times 85.157 \times 10^6)^2}{24\,200}$$

$$= 7.49\,\Omega$$

a value that can be obtained, for example, by forming the antenna from wires of suitable diameter.

$$Q_{2s} = \frac{L_{2s} \cdot \omega c}{R_{2s}} = 56.85$$

and

$$R_{2p} = Q_{2s}^2 \cdot R_{2s} = 24.2\,\text{k}\Omega$$

In this case,

$$Q_{p2\,\max} = \frac{1}{2}\sqrt{\frac{R_{\text{ic}}}{R_{2s}}}$$

$$= \frac{1}{2}\sqrt{\frac{24\,200}{7.49}}$$

$$= 28.42$$

The system formed by the transponder antenna and the integrated circuit has the following values:

$$R_p = R_{\text{ic}} // R_{2p} = 12.1\,\text{k}\Omega$$

and

$$Q_{p2} = \frac{R_p}{L_{2s} \cdot \omega c} = 28.4$$

Now let us look at the transponder antenna and its equivalent series circuit.

This circuit consists of (see *Figure 6.16a*) L_{2s}, R_{2s} and C_{2p} in parallel with the rest, and obviously has an intrinsic resonant frequency and quality factor, whose value clearly depends on the frequency at which the antenna operates.

$$Q_{2s} = \frac{L_{2s} \cdot \omega}{R_{2s}}$$

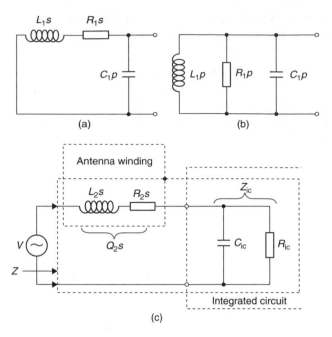

Figure 6.16 Equivalent circuit of the transponder

This series layout of the antenna circuit can be represented as a parallel equivalent circuit consisting of (see *Figure 6.16b*)

$$L_2p = L_2s$$
$$R_2p = Q_2s^2 \cdot R_2s$$
$$C_2p = C_2p$$

whose intrinsic resonant frequency is

$$L_2p \cdot C_2p \cdot \omega c^2 = 1$$

(the same as before, naturally) and whose quality factor can now be written as

$$Q_2p = \frac{R_2p}{L_2s \cdot \omega}$$

and whose value is the same as before, that is,

$$Q_2s = Q_2p \Rightarrow R_2p = Q_2s^2 \cdot R_2s$$

The complete circuit of the system comprising the antenna and the integrated circuit is shown in parallel form in *Figure 6.16c*, in which we can say that

$$L_2p = L_2s$$
$$Cp = C_{ic} + C_{con} + C_2p$$

= the sum of the input capacitances of the integrated circuit, the parallel connection capacitances and the parallel capacitance of the antenna coil

$$Rp = \frac{R_{ic} \cdot R_2 p}{R_{ic} + R_2 p}$$

= the paralleling of the input resistance of the integrated circuit and the equivalent parallel resistance of the antenna coil

$$= 24.2 \, // \, 24.2 = 12.1 \, \text{k}\Omega$$

and

$$L_2 p \cdot Cp \cdot \omega^2 = 1$$

and

$$Qp_2 = \frac{Rp}{L_2 p \cdot \omega}$$

At this stage of the procedure, Qp_2 has been chosen and the values of $L_2 s$ and Cp have been determined.

We can now determine the circuit of the equivalent dipole:

$$Zi = Q_2 s \cdot L_2 s \cdot \omega c$$

$$= 56.84 \times 5 \times 10^{-6} \times 85.157 \times 10^6$$

$$= 24\,200\,\Omega$$

$$= \text{perfect power matching with } R_{ic}$$

on the other hand, the transponder will operate correctly only if

$$V_{lc} = 2.2 \, \text{V}$$

and therefore

$$V_{ic} = Q_2 p \cdot V_{20}$$

therefore

$$V_{20} = \frac{V_{ic}}{Qp_2}$$

$$= \frac{2.2}{28.4}$$

$$= 77.46 \, \text{mV}$$

and

$$V_{20\,e} = V_{20} \cdot Q_2 s$$

$$= 77.46 \times 10^{-3} \times 56.84$$

$$= 4.4 \, \text{V}$$

therefore

$$V_{ic} = \frac{R_{ic}}{Zi + R_{ic}} \cdot V_{20\,e}$$

$$= \frac{24.2}{24.2 + 24.2} \times 4.4$$

$$= 2.2\,\text{V} \quad \text{Q.E.D.}$$

Figure 6.17 summarizes the equivalent circuit of the equivalent dipole.

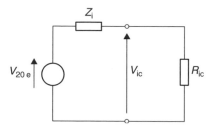

Figure 6.17   Dipole equivalent to the transponder antenna

After this general discussion of the parameters of a transponder fitted with an I·CODE circuit, we shall now examine two specific examples of application:

— the first is the production of either an airline baggage label or a vicinity access card according to ISO 15 693, one or other of these being made in conventional ISO card format (8 cm × 5 cm);

— the second is the production of a flexible label in 5 cm × 5 cm format, for item management (traceability, stock monitoring, etc.).

Important notes on the following examples

(a)   Later on (in Chapter 7) we will see that sometimes, owing to simple technological factors, it is not possible for one or other of the examples to have a sufficient number of turns to provide the same total surface area S_2 of the transponder antenna.

This is because, on the one hand, the value of the tuning capacitance is determined by the integrated circuit, and, on the other hand, there are differences in surface area between an ISO card (8 × 5 cm) and a badge or label (5 × 5 cm); consequently, for a given inductance of the transponder antenna (in our case, $L_2s = 5\,\mu\text{H}$), it is not possible, for technological reasons, to produce identical antennae (in terms of shape, number of turns, thickness of tracks, etc.). We must therefore use, for example, a rectangular antenna in one case and a square one in another, with different numbers of turns.

The result of all this is that, for a given magnetic field strength created by a base station, the two products (ISO card and badge/label) will have slightly different performances in respect of their operating range.

(b)   Within these two types of application, these transponders must be readable equally well in the near and far regions. This is why we set the value of Q_1 to 35.

6.3.7 Baggage label or ISO 15693 vicinity card in the ISO card format

Let us start with the example of an implementation of vicinity contactless transponders in the ISO card format, designed for long-range access card applications (the card being kept in the pocket when passing the base station), or for airline baggage label applications according to IATA specifications, both of these generally having the same physical dimensions (see *Figure 6.18*).

Figure 6.18 Airline baggage label fitted with a 13.56 MHz transponder

Table 6.10 summarizes the physical characteristics proposed for this application.

As regards the electrical parameters, for integrated circuits of the Philips Semiconductors I·CODE type operating at a bit rate of 26 kb/s, we will continue to use the electrical values found in the generic sections above, in other words

Table 6.10

Dimensions of antenna	$= 7.3 \times 4.2$ cm
N_2	$= 6$
S_2	$= (7.3 \times 4.2)$ cm^2
	$= 30.66$ cm^2
$N_2 \cdot s_2$	$= 183.96$ cm^2
Length of one turn	$= 2 \times (7.3 \times 4.2)$ cm
	$= 23$ cm

L_2s	$= 5\,\mu H$
R_2s	$= 7.49\,\Omega$
Q_2s at 13.56 MHz	$= 56.84$
R_2p	$= 24\,200\,\Omega$
Cp	$= 27.6\,pF$
$\omega s_{\text{intrinsic}}$	$= \dfrac{1}{\sqrt{L_2s \cdot Cp_{\text{total}}}}$
	$= \dfrac{1}{\sqrt{5 \times 10^{-6} \times 27.6 \times 10^{-12}}}$
	$= 85.15 \times 10^6\,\text{rd/s}$
$f\,s_{\text{tag proper}}$	$= \dfrac{\omega s}{2\pi}$
	$= \dfrac{85.15 \times 10^6}{6.28}$
	$= 13.56\,\text{MHz}$: the transponder is tuned to the carrier
R_{ic}	$= 24.2\,k\Omega$
R_p	$= 12.1\,k\Omega$
Qp_2	$= 28.42$

Calculating the typical threshold magnetic field strength and induction   The typical threshold field strength H_{dt} is given by the following formula:

$$H_{dt} = \frac{\sqrt{\left[1 - \left(\frac{\omega c}{\omega r}\right)^2\right]^2 + \left[\left(\frac{L_2p \cdot \omega c}{Rp}\right)^2\right]}}{\omega c \cdot \mu_0 \cdot N_2 s_2} V_{\text{ic typ}}$$

Given that the transponder is tuned ($\omega = \omega r$), and that $Qp_2 = (Rp)/(L_2 p \cdot \omega)$, the above equation takes the form

$$H_{dt} = \frac{L_2 p}{\mu_0 \cdot N_2 s_2 Rp} V_{\text{ic typ}} = \frac{1}{\mu_0 \cdot N_2 s_2} \cdot \frac{1}{Qp_2} \cdot \frac{1}{\omega} V_{\text{ic typ}}$$

therefore

$$H_{dt} = \frac{2.2}{4 \times 3.14 \times 10^{-7} \times 6 \times 30.66 \times 10^{-4} \times 28.42 \times 2 \times 3.14 \times 13.56 \times 10^6}$$

$$= \frac{2.2}{59.9}$$

$$= 39.35\,\text{mA/m}$$

and the associated value of B_{dt}

$$B_{dt} = \mu \cdot H_{dt}$$
$$= 4 \times 3.14 \times 10^{-7} \times 39.35 \times 10^{-3}$$
$$= 49.42 \, \text{nT}$$

Calculating the flux captured by the transponder antenna   The flux captured by the transponder antenna is as follows:

$$\Phi_2 = B_{dt} N_2 s_2$$
$$= 49.42 \times 10^{-9} \times 6 \times 30.66 \times 10^{-4}$$
$$= 0.9091 \, \text{nWb/m}^2$$

Calculation of the induced voltage V_{20}

$$V_{20} = (\mu \cdot H_{dt} N_2 s_2) \cdot \omega$$
$$= \Phi \cdot \omega = (0.9091 \times 10^{-9}) \times 85.15 \times 10^6$$
$$= 77.4 \, \text{mV}$$

Calculation of the voltage applied to the integrated circuit V_{ic}

$$V_{ic} = Qp_2 \cdot V_{20}$$
$$= 28.42 \times 77.4 \times 10^{-3}$$
$$= 2.2 \, \text{V}$$

Value of the impedance presented by the transponder   Let us calculate the impedance presented by the whole electrical circuit of the transponder. We know that

$$Z = \sqrt{\frac{Rsc^2 + (Lsc + Rsc R_{ic} Cp)^2 \cdot \omega^2}{1 + R_{ic}{}^2 Cp^2 \cdot \omega^2}} = \sqrt{\frac{Rsc^2 + (Lsc + Rsc R_{ic} Cp)^2 \cdot \omega^2}{1 + \dfrac{R_{ic}^2 Cp}{Lsc}}}$$

$$= \frac{\sqrt{7.49^2 + [5 \times 10^{-6} + (7.49 \times 24.2 \times 10^3 \times 27.6 \times 10^{-12})]^2 \times 7251 \times 10^{12}}}{\sqrt{1 + (24.2^2 \times 10^6 \times 27.6^2 \times 10^{-24} \times 7251 \times 10^{12})}}$$

$$= \frac{\sqrt{56.1 + 725\,100}}{\sqrt{3230}}$$

$$= \frac{851.56}{56.83}$$

$$= 14.98 \, \Omega$$

The quick method described in Chapter 3 would have given us the following (still in the case of a transponder tuned to the carrier frequency):

$$Rs_{tot} = Rsc + \frac{Lsc^2 \cdot \omega^2}{R_{ic}}$$

$$= 7.49 + \frac{(25 \times 10^{-12}) \times (7251 \times 10^{12})}{24.2 \times 10^3}$$

$$= (7.49) + (7.49)$$

$$= 14.98 \, \Omega$$

Leaving aside the note concerning perfect matching between the generator (7.49 Ω) and load (7.49 Ω), the rest requires no comment.

Current flowing in the antenna winding   Now that we know the impedance, we can easily calculate the different values of the currents flowing in the different elements, first of all the current flowing through the transponder antenna coil.

$$I_2 = \frac{V_{20}}{Z}$$

$$= \frac{77.46 \times 10^{-3}}{14.98}$$

$$= 5.17 \, \text{mA}$$

Let us take another look at the value of the current I_2 that we have just calculated. This current is certainly the current flowing through the inductance of the transponder antenna coil and the global tuning capacitance (Cp including C_{ic} and $C_{parasitic}$), since the system is resonating and therefore has an excess current coefficient under load Qp_2 equal to the quality factor (for further information, see the appendices). Clearly, the presence of this current I_2 is due to the existence of an induced current i_2, which itself is directly caused by the phenomenon of magnetic induction, since

$$\Phi_2 = B_{dt} \cdot S_2 = B_{dt} \cdot (N_2 s_2)$$

$$= (49.42 \times 10^{-9} \times 6) \times 30.66 \times 10^{-4}$$

$$= 0.9091 \, \text{nWb/m}^2$$

$$= L_{2}s \cdot i_2$$

The current induced in the unloaded winding of the transponder antenna is therefore

$$i_2 = \frac{\Phi_2}{L_{2}s}$$

$$= \frac{0.9091 \times 10^{-9}}{5 \times 10^{-6}}$$

$$= 0.18 \, \text{mA}$$

and therefore, because of the resonance of the tuned circuit,

$$I_2 = i_2 \cdot Qp_2$$
$$= 0.18 \times 28.4$$
$$= 5.17 \, \text{mA} \quad \text{Q.E.D.}$$

Current distribution   The current flowing in the transponder antenna coil is divided into three branches (see *Figure 6.19a*) having the following effective values:

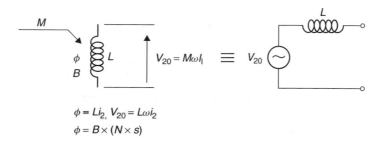

$$\phi = Li_2, \ V_{20} = L\omega i_2$$
$$\phi = B \times (N \times s)$$

Figure 6.19a   Voltage induced in the transponder antenna

(a) $I_{C_{ic}} = V_{20} \cdot (C_{ic} \cdot \omega)$

$$= 2.2 \times (23 \times 10^{-12} \times 85.151 \times 10^6)$$
$$= 4.3 \, \text{mA}$$

(b) $I_{C_{\text{parasitic}}} = V_{20} \cdot C_{\text{par}} \cdot \omega$

$$= 2.2 \times (4.6 \times 10^{-12} \times 85.151 \times 10^6)$$
$$= 0.861 \, \text{mA}$$

(c) $I_{R_{ic}} = \dfrac{2.2}{24.2 \times 10^3}$

$$= 90 \, \mu\text{A}$$

Since the value of $I_{C_{ic}}$ is much greater than the value of $I_{R_{ic}}$, in spite of the 90° phase separation between them (see the phase diagram in *Figure 6.19b*), the effective current actually flowing in the input pin of the integrated circuit I_{ic} (the sum of the currents $I_{C_{ic}}$ and $I_{R_{ic}}$) is therefore approximately 4.3 mA.

Voltage distribution   The potential difference developed across the terminals of the antenna inductance as a result of the excess voltage caused by the tuning of the circuit is

$$V_{L_{2s}} = (L_2 s \cdot \omega) \cdot (i_2 Q p_2)$$
$$= (5 \times 10^{-6} \times 2 \times 3.14 \times 13.56 \times 10^6) \cdot (0.18 \times 10^{-3} \times 28.42)$$
$$= 2.178 \, \text{V}$$

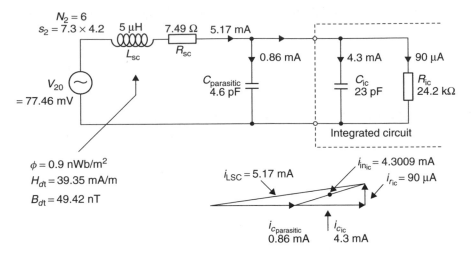

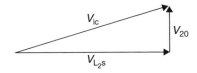

Figure 6.19b   Distribution of currents among the components of the transponder

therefore see *Figure 6.20*

Figure 6.20   Distribution of voltages among the components of the transponder

$$V_{ic} = \sqrt{V_{L_{2s}}{}^2 + V_{20}}$$

$$V_{ic} = 2.2\,\text{V}$$

Power dissipation in the transponder in the presence of the threshold field strength H_d
The power dissipated in the integrated circuit,

$$P_{C_{ic}} = 0.0\,\text{W},$$

is only reactive power, since the current $(i_{C_{ic}})$ and voltage (V_{ic}) are in quadrature:

$$P_{R_{ic}} = \frac{V_{ic}{}^2}{R_{ic}}$$

$$= \frac{2.2^2}{24.2 \times 10^3}$$

$$= 200\,\mu\text{W}$$

and this corresponds to the actual dissipated power in watts:

$$P_{\text{tot}_{ic}} = P_{C_{ic}} + P_{R_{ic}} = 200\,\mu\text{W}$$

Final note Very often, in order to achieve secure operation regardless of the tolerances of the components (L_1s, L_2s, Cp, etc.), users decide to operate their systems with a typical magnetic field strength H_{typ} of the order of twice the threshold field strength H_{dt}, that is, 80 mA/m in our example.

Since all the values calculated in the above sections are proportional to the magnetic field strength, all the resulting values must be multiplied by two, that is,

$B_d = 100.5\,\text{nT}$

$\Phi_d = 1.848\,\text{nW/m}^2$

$i_2 = 0.36\,\text{mA}$

which gives us

$$V_{L_2s} = 4.356\,\text{V}$$

$$V_{20} = 154.8\,\text{mV}$$

and therefore

$V_{ic} = 4.4\,\text{V}$ disregarding the regulation by the shunt circuit of the controller.

Let us continue with the example considered in the preceding sections, with the values summarized in *Table 6.11*.

Calculating the typical coupling coefficient k_{typ} of the application Continuing with our example, the voltage applied to the base station antenna is as follows:

$$V_1 = (L_2s \cdot \omega)I_1$$
$$= (1.8 \times 10^{-6} \times 85.157 \times 10^6) \times 0.79$$
$$= 122.4\,\text{V}$$

We can then calculate k_{typ} required for the application to operate correctly:

$$V_{20\,typ} = V_1 \cdot k_{typ}\sqrt{\frac{L_2}{L_1}}$$

$$k_{typ} = \frac{V_{20\,typ}}{V_1\sqrt{\dfrac{L_2}{L_1}}}$$

$$= \frac{77.4 \times 10^{-3}}{122.4\sqrt{\dfrac{5}{1.8}}}$$

$$= 0.379 \times 10^{-3}$$

$$= 0.038\%$$

Table 6.11

**Simultaneous long- and short-range application
base station end**

Antenna

L_1s	$= 1.8\,\mu H$
Q_1	$= 35$ near and far reading/writing
N_1	$= 1$
I_1	$= 0.79\,A$
V_1	$= 122.4\,V$ rms
r_1	$= 33.4\,cm$
H_0	$= 1.18\,A/m$
B_0	$= 1.5\,\mu T$
Pa	$= 10\,mW$

Transponder end

Antenna only

L_2s	$= 5\,\mu H$
R_2s	$= 7.49\,\Omega$
Q_2s	$= 56.8$
therefore R_2p	$= 24.2\,k\Omega$
N_2	$= 6$
s_2	$= 7.3 \times 4.2\,cm$
	$= 30.66\,cm^2$
S_2	$= 183.96$

Integrated circuit only

V_{ic}	$= 2.2\,V$ rms
P_{typ}	$= \dfrac{V_{ic}^2}{R_{ic}}$
	$= 200\,\mu W$
therefore R_{ic}	$= \dfrac{V_{ic}^2}{P_{max}}$
	$= 24.2\,k\Omega$

"Antenna + integrated circuit" system

Rp	$= R_2p \,//\, R_{ic} = 12.1\,k\Omega$
Qp_2	$= 28.42$

This value enables us to calculate the theoretical typical operating range, and also the mutual inductance of the system.

Typical minimum and maximum operating range   The typical operating range is found by means of the formula

$$k = \left(\mu \frac{r^2}{2(r^2 + d^2)^{3/2}} \right) \cdot N_1 N_2 s_2 \frac{1}{\sqrt{L_1s \cdot L_2s}}$$

where

$$k = 3.8 \times 10^{-4} \quad \mu = 4 \times \pi \times 10^{-7} \quad r = 0.35\,\text{m}$$
$$N_1 = 1 \quad N_2 = 6 \quad s_2 = 30.66\,\text{cm}^2$$
$$L_1 s = 1.8\,\mu\text{H} \quad L_1 s = 5\,\mu\text{H}$$

we obtain $d_{\text{max typ}} = 1.01$ disregarding the tolerances and dispersions of the different elements.

Mutual inductance   In this case, the mutual inductance is as follows:

$$M_{\text{typ}} = k\sqrt{L_2 s \cdot L_1 s}$$
$$= 3.8 \times 10^{-4}\sqrt{1.8 \times 10^{-6} \times 5 \times 10^{-6}}$$
$$= 1.4\,\text{nH}$$

Verification:

$$V_{20} = \frac{M \cdot \omega c}{\sqrt{\left[\left(1 - \left(\frac{\omega c}{\omega r}\right)^2\right)^2 + \left(\frac{L_2 s \cdot \omega c}{R_2}\right)^2\right]}} \cdot I_1$$

Given that the transponder is tuned to the carrier ($\omega = \omega r$),

$$I_1 = \frac{V_{20}}{M \cdot \omega c}$$
$$= \frac{77.4 \times 10^{-3}}{1.14 \times 10^{-9} \times 85.157 \times 10^6}$$
$$= 797\,\text{mA rms}$$
$$= 0.79\,\text{A} = \text{OK!}$$

Return voltage induced in the base station antenna   To complete this application, let us assign an order of magnitude to the voltage induced in the base station antenna during the stage of load modulation of the transponder used to provide the downlink communication from the transponder to the base station. As mentioned in Chapter 3, by reciprocity, this voltage is

$$\Delta V_1 = \Delta V_2 \cdot k\sqrt{\frac{L_1}{L_2}} \cdot Q_1$$

where ΔV_2 represents the voltage variation produced by the electronic circuit of the transponder load modulator across the terminals of the antenna winding. This variation is usually of the order of 3 to 5 V peak to peak (shown as V pp) in commercially available circuits.

Knowing this value, in the most difficult case – in other words, when the coupling coefficient is lowest – we can put a number on the variation of voltage ΔV_1 developed across the terminals of the base station antenna, which must be extracted from the noise, amplified, filtered, demodulated and decoded. Let us return to the values used before, with

$L_1 = 1.8\,\mu H$

$L_2 = 5\,\mu H$

$Q_1 = 35$ (chosen for applications using transponders operating simultaneously at short and long range), and

$k \;\;= 0.38 \times 10^{-3}$

and assume that $\Delta V_3 = 3$ V pp (square signal)

We find

$$\Delta V_1 = \Delta V_2 \cdot k \sqrt{\frac{L_1}{L_2}} \cdot Q_1$$

$$= 3 \times 0.38 \times 10^{-3} \sqrt{\frac{1.8}{5}} \times 35$$

$$= 23.9\,\text{mV pp (square signal)}$$

If we assume that, when received at the base station, the signal is filtered and only its first harmonic is left, we can estimate the effective value of the received signal, namely,

$$\Delta V_{1\,\text{rms}} = \frac{\dfrac{\Delta V_{1\,\text{pp}}}{2}}{\sqrt{2}} \cdot \frac{4}{\pi}$$

$$= \frac{\dfrac{23.9 \times 10^{-3}}{2}}{1.414} \cdot \frac{4}{\pi}$$

$$= 10.8\,\text{mV rms}$$

Assuming that an effective voltage of 122.4 V is present at the base station antenna during the response phase, as before, we can calculate the voltage ratio "return signal/transmitted signal":

$$\frac{\text{return signal}}{\text{transmitted signal}} = \frac{10.8 \times 10^{-3}}{122.4} = 8.8 \times 10^{-5}$$

that is, in dB:

$$20 \log(8.8 \times 10^{-5}) = -81.16\,\text{dB}$$

NOTE

The aim of this note is to remind you that, for proximity applications, we found a ratio of approximately -50 dB, as against approximately -80 dB for vicinity applications. Only 30 dB apart... but enough to cause plenty of problems!

To summarize, for baggage label or ISO 15 693 vicinity card applications, *see Table 6.12* and *Figure 6.21*.

Table 6.12

Optimization of the base station and transponder parameters for a maximum radiated power of 10 mW and simultaneous short and long-range application

$Pa \quad = 10\,\text{mW}$

Base station antenna

$L_1 s \ = 1.8\,\mu\text{H}$
$R_1 s \ = 0.9\,\Omega$
$N_1 \ = 1$
$D_1 \ = 66.8\,\text{cm}$
$r_1 \ = 33.4\,\text{cm}$
$Q_1 \ = 35$ under load, simultaneous short and long- range operation
$Ra \ = 15.94\,\text{m}\Omega$

$P_1 \ = 2.73\,\text{W}$ because $Q_1 = 35$ instead of 100
$I_1 \ = 0.79\,\text{A}$
$V_1 \ = 122.6\,\text{V}$ rms
$H_0 \ = 1.18\,\text{A/m}$
$B_0 \ = 1.5\,\mu\text{T}$
$H_{66.8} = 100\,\text{mA/m}$

Transponder end

Antenna only

$$L_2 s = 5\,\mu\text{F}$$
$$R_2 s = 7.49\,\Omega$$
$$Q_2 s = 56.8$$

therefore $R_2 p = 24.2\,\text{k}\Omega$

$$N_2 = 6$$
$$s_2 = 7.3 \times 4.2\,\text{cm}$$
$$= 30.66\,\text{cm}^2$$
$$S_2 = 183.96\,\text{cm}^2$$

Integrated circuit only

$$V_{ic} = 2.2\,\text{V rms}$$

$$P_{typ} = \frac{V_{ic}^2}{R_{ic}}$$
$$= 200\,\mu\text{W}$$

therefore $R_{ic} = \dfrac{V_{ic}^2}{P_{max}}$

$$= 24.2\,\text{k}\Omega$$

"Antenna + integrated circuit" system, wholly tuned to the carrier

$Rp \ = R_2 p // R_{ic} = 12.1\,\text{k}\Omega$
$Qp_2 = 28.42$
$H_{dt} \ = 39.35\,\text{mA/n}$
$B_{dt} \ = 49.42\,\text{nT}$

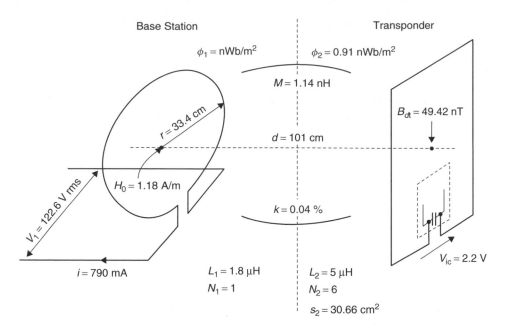

Figure 6.21 Summary of the "base station and associated transponder" system

6.3.8 ISO 15 693 consumable product labels ("smart labels")

These disposable labels are usually specified for applications generally operating at a distance of approximately 60 to 70 cm, with a single antenna, and at a distance of 1 to 1.5 m with what are known as "gate" antennae (as used at the exits from large stores, see *Figure 6.22*). They are generally made in a square physical format, measuring approximately 5×5 cm, and have seven turns each (see *Figure 6.23*).

The physical characteristics of the label antenna are generally as follows:

$N_2 = 7$

$s_2 = (4.5 \times 4.5)\, \text{cm}^2$ (maximum surface area of the transponder antenna)

   $= 20.25\, \text{cm}^2$

$S_2 = N_2 s_2 = 141.75\, \text{cm}^2$

Now let us take a quick look at the induction, field strength and flux characteristics that they require.

Calculating B_{dt} and H_{dt} where $\omega r = \omega c$ Knowing that the operating threshold voltage of the I·CODE integrated circuit is $V_{ic\,typ} = 2.2\,\text{V}$ rms and the optimal value of the quality factor $Qp_2 = 28.4$, we can calculate the threshold magnetic induction required for the label when its tuning frequency is locked to the carrier frequency

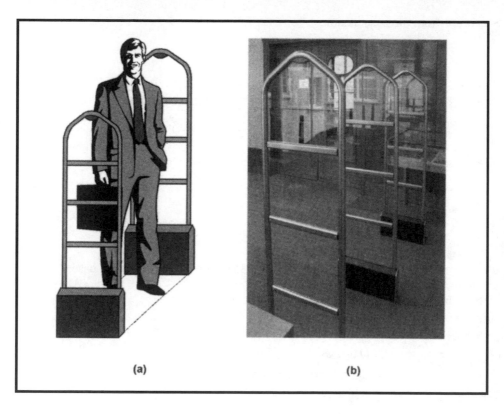

Figure 6.22 Example of exit gates in a store

Figure 6.23 Example of 5×5 cm disposable label

received from the base station:

$$B_{dt} = |V_{\text{ic min}}| \frac{1}{Qp_2 \cdot \omega \cdot N_2 s_2}$$

$$= 2.2 \frac{1}{28.4 \times 85.15 \times 10^6 \times (7 \times 20.25 \times 10^{-4})}$$

$$= 64.3 \, \text{nT, as against } 49.42 \, \text{nT in the preceding case}$$

The associated magnetic field strength is therefore

$$H_{dt} = \frac{B_{dt}}{\mu}$$

$$= \frac{64.3 \times 10^{-9}}{4 \times 3.14 \times 10^{-7}}$$

$$= 51.2 \, \text{mA/m, as against } 39.35 \, \text{mA/m in the preceding case}$$

From these two values we can derive V_{20} and V_{ic}, thus

$$V_{20} = \Phi_2 \cdot \omega$$

$$= (B_{dt} N_2 s_2) \omega$$

$$= (64.3 \times 10^{-9} \times 7 \times 20.25 \times 10^{-4}) \times 85.15 \times 10^6$$

$$= 77.6 \, \text{mV}$$

and

$$V_{\text{ic}} = Qp_2 \cdot V_{20}$$

$$= 28.4 \times 77.6 \, \text{mV}$$

$$= 2.2 \, \text{V} \qquad \text{Q.E.D.}$$

If the induced voltage $V_{20} = 77.6 \, \text{mV}$ has been created, this is because the flux Φ_2 captured by the transponder antenna was

$$\Phi_2 = B_{dt}(N_2 s_2)$$

$$= 64.3 \times 10^{-9} \times (7 \times 20.25 \times 10^{-4})$$

$$= 0.91 \, \text{nWb/m}^2$$

and the current induced without load, i_2, flowing in L_2 would be as follows:

$$\Phi_2 = L_2 s \cdot i_2$$

$$= 5 \times 10^{-6} \times i_2$$

$$i_2 = \frac{0.91 \times 10^{-9}}{5 \times 10^{-6}}$$

$$= 0.182 \, \text{mA}$$

and the current I_2 flowing in the tuned circuit in resonance (see *Figure 6.24*) is

$$I_2 = i_2 \cdot Qp_2$$
$$= 0.182 \times 10^{-3} \times 28.4$$
$$= 5.16 \, \text{mA}$$

In resonance, the impedance of the transponder can be written thus:

$$V_{20} = Z \cdot i_2$$

therefore

$$Z = \frac{77.6 \times 10^{-3}}{0.182 \times 10^{-3}}$$
$$= 426 \, \Omega$$
$$= Qp_2(7.49 + 7.49) = 28.4 \times 15$$
$$= 426 \, \Omega \quad \text{Q.E.D.}$$

NOTE

You should note that, although the integrated circuit used in the preceding example is strictly subject to the same electrical operating conditions, the difference between the physical formats of the antennae of the two types of product – ISO labels on one hand and smart labels on the other hand – causes a change in the magnetic characteristics of the transponders (see the table below), and therefore the operating ranges must in principle be different for the same use of a base station.

	ISO format label	5×5 cm smart label
cm^2	$(6 \times 7.3 \times 4.2) = 183.96$	$(7 \times 4.5 \times 4.5) = 141.75$
mA/m	39.35	51.2
nT	49.42	64.3

First sub-example Since we now know that, when the base station and transponder are perfectly matched, the values of B_{dt} (64.3 nT), H_{dt} (51.2 mA/m) and Φ_{dt} (0.9 nWb/m^2) required by this type of label to operate correctly, regardless of the operating range (... values which, incidentally, should always be stated by the manufacturer of finished products such as transponders or "inlets"), let us estimate, by way of a first example, the characteristics which the base station must have to ensure that these labels (transponders) operate correctly at 60 cm. In this case, we must write

$$B_{dt} = B_{60} = 64.3 \, \text{nT}$$

Assuming that the base station antenna has the same physical structure as that described previously – in other words, a radius r_1 of 20 cm or a rectangle measuring approximately 30 cm $\times$ 40 cm – we can write

$$a = \frac{d}{r_1} = \frac{60}{20} = 3$$

from which we derive the correction factor between B_0 and B_{60}.

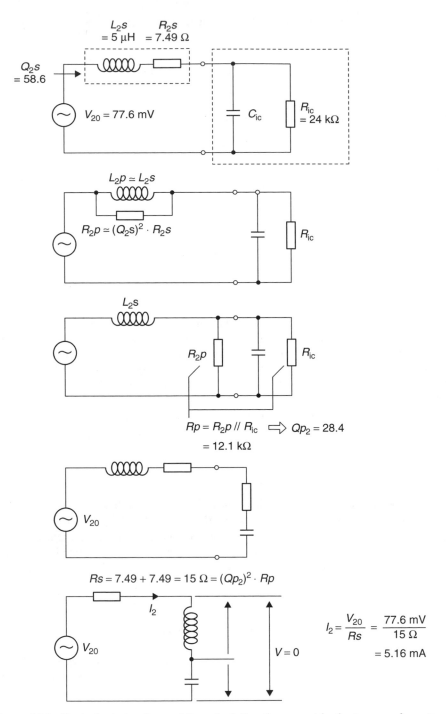

Figure 6.24   Equivalent circuits used for calculating the current in the transponder antenna

Since $a = 3$, then $(1 + a^2)^{3/2} = 31.62$, that is,

$$B_0 = 31.62 \ B_{60}$$
$$= 31.62 \times 64.3 \times 10^{-9}$$
$$= 2.03 \, \mu\text{T}$$

This means that, in order to meet the application requirements, the magnetic field in the centre ($d = 0 \, \text{cm}$) of the base station antenna must have a strength of

$$H_0 = \frac{B_0}{\mu_0}$$
$$= \frac{2.03 \times 10^{-6}}{4 \times \pi \times 10^{-7}}$$
$$= 1.62 \, \text{A/m}$$

On the other hand, given that

$$H_0 = \frac{N_1 I_1}{2r_1}$$

this can be used to derive the value of the current which must flow through the antenna ($N_1 = 1$ and $r_1 = 30 \, \text{cm}$), that is,

$$I_1 = \frac{1.62 \times 2 \times 0.2}{1} = 0.648 \, \text{A}$$

and, given that

$$I_1 = \sqrt{\frac{P_1 Q_1}{L_1 s \cdot \omega}}$$

and therefore

$$P_1 = \frac{(L_1 s \cdot \omega) I_1^2}{Q_1}$$
$$= \frac{(1.8 \times 10^{-6}) \times (85.15 \times 10^6) \times 0.648^2}{35}$$
$$= 1.84 \, \text{W}$$

and on the other hand the associated power Pa radiated by the base station antenna,

$$Pa = Ra \cdot I_1^2$$
$$= 0.13(N_1 \cdot \pi r_1^2)^2 I_1^2$$
$$= 0.13 \times (1 \times 3.14 \times 0.2^2)^2 \times 0.648^2$$
$$= 0.860 \, \text{mW}$$

Second sub-example   At the start of this book (Chapter 1), we saw that, if we wish to operate at a known distance, there is a special radius r_1 ($d = 1.414 r_1$) at which the current flowing through the antenna is optimized to provide the induction required for the application. To compare performance, we shall start from this second assumption.

Since $d = 60\,cm$ and $a = d/r_1 = 1.414$, we obtain the value of $r_1 = 42.43\,cm$ (giving a total surface area of the antenna $S_1 = 5644\,cm^2$, for example, in the form of a rectangular arch measuring $40\,cm \times 1.40\,cm$).

Knowing that $a = 1.414$, we can find the correction factor $(1 + a^2)^{3/2} = 5.86$ between B_0 and B_{60}, that is,

$$
\begin{aligned}
B_0 &= B_{60} \times 5.86 \\
&= 64.3 \times 10^{-9} \times 5.86 \\
&= 0.377\,\mu T
\end{aligned}
$$

This means that, in order to meet the application requirements, the magnetic field in the centre ($d = 0\,cm$) of the base station antenna must have a strength of

$$
\begin{aligned}
H_0 &= \frac{B_0}{\mu_0} \\
&= \frac{0.377 \times 10^{-6}}{4\pi \times 10^{-7}} \\
&= 0.3\,A/m
\end{aligned}
$$

On the other hand, given that

$$
H_0 = \frac{N_1 s_1}{2r_1}
$$

this can be used to derive the value of the current which must flow through the antenna ($N_1 = 1$ and $r_1 = 30\,cm$), that is,

$$
I_1 = \frac{0.3 \times 2 \times 0.424}{1} = 0.254\,A
$$

and on the other hand, given that

$$
I_1 = \sqrt{\frac{P_1 Q_1}{L_1 s \cdot \omega}}
$$

therefore

$$
\begin{aligned}
P_1 &= \frac{L_1 s \cdot \omega \cdot I_1^2}{Q_1} \\
&= \frac{(1.8 \times 10^{-6})(85.15 \times 10^6) \times 0.254^2}{35} \\
&= 0.28\,W
\end{aligned}
$$

On the other hand, the associated power Pa radiated by the base station antenna is

$$
\begin{aligned}
Pa &= Ra \cdot I_1^2 \\
&= 0.13(N_1 \cdot \pi \cdot r_1^2)^2 I_1^2
\end{aligned}
$$

$$= 0.13 \times (1 \times 3.14 \times 0.424^2)^2 \times 0.254^2$$

$$= 2.67 \, \text{mW}$$

This solution has the advantage of greatly reducing the current flowing through the base station antenna. The price to be paid for this decrease in current is the considerable increase in the physical size of the antenna, although sometimes this is not a problem (e.g. in the exit gate of a store).

Now it is up to you to choose, according to your proposed application: everyone has their own problems!

Third sub-example Now let us specify, for example, a maximum radiated power of $Pa = 10 \, \text{mW}$ rms with the same values of

$B_0 \quad = 1500 \, \text{nT rms}$

$\text{const} = 0.13$

$N_1 \quad = 1$

In this case, we can calculate

$$H_0 = \frac{B_0}{4\pi \times 10^{-7}}$$

$$= 1.2 \, \text{A/m rms}$$

Given that

$$Pa = \frac{\text{const} \cdot \pi^2 N_1{}^6 \cdot I_1{}^6}{(2 \times H_0)^4}$$

therefore

$$I_1{}^6 = \frac{Pa \cdot (2H_0)^4}{\text{const} \cdot \pi^2 N_1{}^6}$$

and therefore

$$I_1 = 0.8 \, \text{A rms}$$

enabling us to evaluate r_1:

$$r_1 = \frac{N_1 I_1}{2 \times H_0}$$

and therefore $r_1 = 33.34 \, \text{cm}$, that is, the diameter is $66.7 \, \text{cm}$.

Since labels in the physical format have a threshold magnetic induction of B_{dt} equal to $64.3 \, \text{nT}$, we can calculate the distance at which the label can operate:

$$\frac{B_0}{B_{dt}} = \frac{1500}{64.3}$$

$$= 23.33$$

$$= (1 + a^2)^{3/2}$$

and therefore

$$a = 2.7$$

so the operating range is

$$d_{\text{max typ}} = a \cdot r_1$$
$$= 2.7 \times 33.4$$
$$= 90\,\text{cm}$$

Verification:

Surface area of one turn:

$$s_1 = \pi r_1{}^2$$
$$= 3.14 \times 0.3334^2$$
$$= 3490\,\text{cm}^2, \text{ that is, a rectangle measuring } 70 \times 50\,\text{cm}$$
$$N_1 = 1$$
$$S_1 = N_1 s_1$$

and therefore

$$Pa = \text{const} \cdot S_1{}^2 \cdot I_1{}^2$$
$$= 0.13 \times (0.1212) \times 0.8^2$$
$$= 10\,\text{mW}$$

If the value of $L_1 s$ is $1.8\,\mu\text{H}$, the voltage across the terminals of the base station antenna will be

$$U_{Ls} = Z_{L_1 s} \cdot I_1 = (L_1 s \cdot \omega) I_1$$
$$= (1.8 \times 10^{-6} \times 85.157 \times 10^6) \times 0.8$$
$$= 122.6\,\text{V rms}$$

In this case, the magnetic flux produced by the base station antenna will be

$$\Phi_1 = L_1 s \cdot I_1$$
$$= 1.8 \times 10^{-6} \times 0.8$$
$$= 1440\,\text{nWb/m}^2$$

Example of the SL EV900 demonstration kit supplied by Philips Semiconductor

The characteristics of this demonstration kit are as follows:

— The base station antenna is rectangular and measures $40\,\text{cm} \times 30\,\text{cm} = 1200\,\text{cm}^2$. As a first approximation, this antenna is practically equivalent to a circular antenna with the same area, having an equivalent radius of approximately $20\,\text{cm}$ ($19.5\,\text{cm}$ to be precise) or a diameter of approximately $40\,\text{cm}$.

— The inductance of the base station antenna is $L_1 s = 1.2\,\mu\text{H}$ (between $500\,\text{nH}$ and $4\,\mu\text{H}$) and its quality factor Q_1 is 35, providing better communication at short range.

— The nominal threshold characteristics required for the correct operation of an I·CODE label with a surface area of $5 \times 5\,\mathrm{cm}$ (antenna with 7 turns measuring $4.5 \times 4.5\,\mathrm{cm}$) are, as demonstrated,

$$H_{dt} = 51.2\,\mathrm{mA/m}$$

$$B_{dt} = 64.3\,\mathrm{nT}$$

If we wish to operate at $60\,\mathrm{cm}$,

$$H_{dt} = H_{60}$$

Therefore, for $a = d/r = 60/19.5 = 3.08$, in other words, a correction factor of $(1 + a^2)^{3/2} = 33.95$, that is,

$$H_0 = 33.95 H_{60} = 1.738\,\mathrm{A/m}$$

given that

$$H_0 = \frac{N_1 I_1}{2r}$$

For the current I_1 which must flow through the antenna ($N_1 = 1$ and $r = 20\,\mathrm{cm}$), we obtain

$$I_1 = \frac{1.738 \times 2 \times 0.2}{1} = 0.695\,\mathrm{A}$$

$$= \sqrt{\frac{P_1 Q_1}{L_1 s \cdot \omega}}$$

and therefore

$$P_1 = \frac{L_1 s \cdot \omega \cdot I_1^2}{Q_1}$$

$$= \frac{(1.2 \times 10^{-6})(85.15 \times 10^6) \times 0.695^2}{35}$$

$$= 1.4\,\mathrm{W}$$

The radiation resistance Ra is as follows:

$$Ra = 0.135^2$$

$$= 0.13 \times 0.1256^2$$

$$= 2.05\,\mathrm{m\Omega}$$

and the radiated power Pa is

$$Pa = Ra \cdot I_1^2$$

$$= 2 \times 10^{-3} \times 0.695^2$$

$$= 0.966\,\mathrm{mW}$$

6.4 Applications and Conformity with Standards

In spite of the use of the terms *proximity* and *vicinity*, the standards relating to contactless devices (ISO 14 443, ISO 15 693 and ISO 18 000) do not actually specify the operating distances of a system. In the industry, it is generally accepted that "proximity" means an operating distance of 8 to 10 cm and "vicinity" means a distance of the order of 60 to 70 cm.

The documents may not deal with distances, but they provide detailed specifications of the required binary coding, bit rate, types of modulation, and so on,[1] and state that, to conform to the standard, transponders "must operate continuously" in the following ranges of values:

Magnetic field strength:

H (in A/m)	ISO 14 443	7.5		1.5	
	ISO 15 693		5		0.15

Translated into magnetic induction in air, as follows:

B (in nT)	ISO 14 443	9400		1884	
	ISO 15 693		6280		188.4

It is therefore up to you to associate the maximum values of the distances at which your system is to operate with the standards, and thus to design a base station capable of delivering the necessary magnetic field strength.

However, I do not want to leave my readers in such an awkward position, so the following sections will assist you in this very difficult mission.

6.4.1 The case of the ISO 14 443 proximity standard

If we follow this standard to the letter, then a contactless card must be capable of operating correctly from 1.5 to 7.5 A/m if it is to be called a "proximity" card. Here is a rather crude example according to the standard.

(1) First of all, assume that a base station having a circular antenna, with a radius of r_1, according to ISO 14 443, produces at its centre a maximum magnetic field of 7.5 A/m – thus ensuring that no card is ever unexpectedly destroyed.

(2) Returning to the general magnetic field equation $H = f(d)$, we know that the ratio $(H_0/H_{dt} = 7.5\,\text{A/m})/(1.5\,\text{A/m}) = 5$ is equal to $(1 + a^2)^{3/2}$ so that we can immediately determine a, thus $a = 1.39$. Knowing the value of a, we can determine the maximum possible operating distance d on the principal axis of the antenna, since, by definition, we have assumed that $a = d/r_1$.

Therefore, if the system is to be called a "proximity" system, according to ISO 14 443, with assured operation at a distance d of $d = 10$ cm (the distance corresponding

---

[1] For further details, see the author's previous book on RFID.

to at least the weakest field of 1.5 A/m), the circular base station antenna must have a radius $r_1 = d/a = 10/1.39 = 7.2$ cm, in other words, a diameter of 14.4 cm. The surface area of the base station antenna will therefore be

$$N_1 = 1$$
$$S_1 = \pi r_1^2$$
$$= 3.14 \times 7.2^2$$
$$= 162.8 \, \text{cm}^2$$

(or an equivalent rectangle measuring 16.3×10 cm or a square measuring 13×13 cm) and its radiation resistance Ra will be as follows:

$$Ra = \text{const} \cdot S_1^2$$
$$= 0.13 \times (162.8 \times 10^{-4})^2$$
$$= 34.45 \, \mu\Omega$$

We also know that

$$H_0 = \frac{N_1 I_1}{2r_1}$$

and therefore, in order to produce a magnetic field strength $H_{0\,\text{max}}$ of 7.5 A/m, the base station antenna must be subject to a current I_1:

$$I_{1\,\text{max}} = 7.5 \times (2 \times 0.72)$$
$$= 1.08 \, \text{A rms}$$

This last value can be used to determine the power radiated by the base station antenna:

$$Pa_{\text{max}} = Ra \cdot I_1^2$$
$$= 34.45 \times 10^{-6} \times 1.08^2$$
$$= 40.2 \, \mu\text{W}$$

NOTE

At the start of this section, I described this example as "crude" because our approach was based on the fact that, in order to operate correctly at 10 cm, the proposed contactless card requires a magnetic field strength of 1.5 A/m – in other words, it would be a card which is barely acceptable according to the standard, being the least sensitive and thus the "worst" card on the market. Fortunately, this case would never be present in the industrial setting – except for a very low-cost application.

We have also seen in this chapter that the power in watts, P_1, is equal to

$$P_1 = \frac{L_1 \cdot \omega c}{N_1^2 \cdot Q_1} \sqrt[3]{\frac{Pa(2 \times H_0)^4}{\text{const} \cdot \pi^2}} = f(Pa)$$

Retaining the numerical values used in the previous examples relating to proximity applications,

$L_1 = 500\,\text{nH}$

$Q_1 = 35$

$N_1 = 1$

we find

$$P_1 = \frac{500 \times 10^{-9} \times 85.157 \times 10^6}{35} \sqrt[3]{\frac{40.2 \times 10^{-6} \times (2 \times 7.5)^4}{0.13 \times 3.14^2}}$$

$$= 1.216 \times 1.156 = 1.4\,\text{W}$$

We shall not worry about the absolute value of this power for the time being, but move on to another aspect which is important for global compliance with the standards.

Constraints imposed by ETSI 300 330 FCC 47, part 15   Apart from the functional aspects covered by ISO 14 443, in Europe we must also comply with ETSI 300 330 governing the level (amplitude and limit) of radio frequency pollution (see *Figure 6.25a*) which or more or less the same as FCC 47 part 15.

In the case of proximity systems according to ISO 14 443, as shown in my previous book, apart from the principal lobe, the level of radiation of the sidebands is the decisive factor in conformity (or non-conformity) with the current ETSI 300 330 standards, for which the values are measured in the far field at 10 m.

Since ISO 14 443 specifies the bit rate, bit coding and type of modulation (100% ASK or 8%–14% max. ASK) and the collision management for the uplink (from the base station to the transponder), the spectrum (in terms of shape and amplitude, relative levels of carrier frequency/sidebands, etc.) of the baseband signal transposed after RF modulation is very well known, and must lie within the limits of the standard (see *Figure 6.25b*).

In the case of a radiated power of $40.2\,\mu\text{W}$ in the carrier, let us now see whether the spectrum of the radiated signal lies within the limits of ETSI 300 330.

In a previous book I showed that, in the "far field", for a circular antenna, the electrical field strength is

$$E \approx \frac{7\sqrt{Pa}}{d}$$

In our example and in the ETSI standard measurement conditions, at 10 m we find

$$E \approx \frac{7\sqrt{40.2 \times 10^{-6}}}{10}$$

$$\approx 4.43\,\text{mV/m}$$

Converting this value to dB μv/m, we obtain

$$E \approx 20\log(4430)$$

$$\approx 72.94\,\text{dB}\,\mu\text{V/m}$$

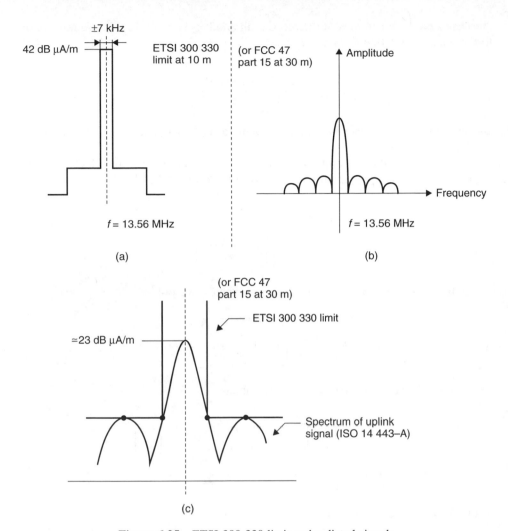

Figure 6.25 ETSI 300 330 limit and radiated signals

and this is then converted to magnetic field strength

$$H \, \mathrm{dB} \, \mu A/m \approx (E \text{ in } \mathrm{dB} \, \mu V/m) - 51$$

$$\approx 21.44 \, \mathrm{dB} \, \mu A/m$$

Now, allowing for the shape of the spectrum radiated by an ISO 14 443 – A signal, *Figure 6.25c* shows that the carrier can be as much as 23 dB $\mu A/m$ while still conforming to ETSI 300 330, and therefore that our proposed proximity solution (10 cm) is viable "on paper".

It will be up to the designer to check, for example, that the output stage of the base station that he has designed, associated with the filter elements provided, does not generate supplementary radiation due to any non-linearity of the driving signal.

Another view   We can look at the problem in another way. To do this, we must return to the equation relating P_1 and Pa:

$$P_1 = \frac{L_1 \cdot \omega c}{N_1{}^2 \cdot Q_1} \sqrt[3]{\frac{Pa(2 \times H_0)^4}{\text{const} \cdot \pi^2}} = f(Pa)$$

and assume that, whatever happens, we want to have magnetic field strength $H_{10} = 1.5$ A/m at $d = 10$ cm.

Now let us find the value of the radius of the base station antenna which meets the requirements of the application while conforming to the ETSI limits.

We know that

$$H_{10} = \frac{H_0}{\sqrt{\left[1 + \left(\dfrac{d}{r}\right)^2\right]^3}}$$

Replacing H_{10} and d with their respective values, we obtain

$$H_0 = 1.5 \sqrt{\left[1 + \left(\frac{0.1}{r}\right)^2\right]^3} = f(r)$$

Bringing H_0 back into the equation, we obtain

$$P_1 = \frac{L_1 \cdot \omega c}{N_1{}^2 \cdot Q_1} \sqrt[3]{\frac{Pa\left[2 \times 1.5 \times \left(1 + \left(\dfrac{0.1}{r}\right)^2\right)^{3/2}\right]^4}{\text{const} \cdot \pi^2}} = f(r)$$

On the other hand, given that

$$E_{\text{ETSI 10 m}} = \frac{7\sqrt{pa}}{d = 10\,\text{m}}$$

and therefore

$$Pa = \left(\frac{E_{\text{ETSI 10 m}} \times 10}{7}\right)^2$$

we can bring Pa into the equation for P_1 to find the value of r:

$$P_1 = \frac{L_1 \cdot \omega c}{N_1{}^2 \cdot Q_1} \sqrt[3]{\frac{\left(\dfrac{E_{\text{ETSI 10 m}} \times 10}{7}\right)^2 \cdot \left[2 \times 1.5 \times \left(1 + \left(\dfrac{0.1}{r}\right)^2\right)^{3/2}\right]^4}{\text{const} \cdot \pi^2}} = f(r)$$

Example

Returning to the values used previously:

$L_1 = 500\,\text{nH}$

$N_1 = 1$

$Q_1 = 35$

and given that, to conform to ETSI 300 330, we must not exceed the maximum field strength H, measured at $10\,\text{m}$,

$$H_{\text{ETSI}\,10\,\text{m}}(\text{in dB}\,\mu\text{A/m}) \approx 23\,\mu\text{A/m}$$

we can comply with ETSI 300 330 by specifying

$$E_{\text{ETSI}\,10\,\text{m}}(\text{in dB}\,\mu\text{V/m}) = H\ \text{in dB}\,\mu\text{A/m} + 51.5$$

$$= 74.5\ \text{in dB}\,\mu\text{V/m}$$

or alternatively

$$E_{\text{ETSI}\,10\,\text{m}} = 5.31\,\text{mV}$$

$$P_1 = 1.216 \sqrt[3]{\dfrac{57.53 \times 10^{-6} \times 81 \left(1 + \left(\dfrac{0.1}{r}\right)^2\right)^6}{1.28}}$$

$$= 1.216 \sqrt[3]{4660 \times 10^{-6}\left[1 + \left(\dfrac{0.1}{r}\right)^2\right]^6}$$

$$= 1.216 \times 16.7 \times 10^{-2}\left[1 + \left(\dfrac{0.1}{r}\right)^2\right]^2$$

$$= 0.203\left[1 + \left(\dfrac{0.1}{r}\right)^2\right]^2 = f(r)$$

NOTE

The example described here is based on an ISO 14443 – A application using 100% ASK modulation, and therefore with a low "carrier/sideband" ratio, which some people might consider to be "unfavourable" by comparison with the 10% ASK modulation used according to ISO 14443 – B.

In fact, although it might appear otherwise on a first approximation, this is not so, since Part B of the standard indicates that the maximum modulation index that can be taken into account is 14%. In this case, if the bit coding, bit rate and the type of management collision according to type B are accepted, then because of the higher ratio between the carrier and the sidebands this enables the carrier to be increased, while conforming to the ETSI standard, by approximately $14\,\text{dB}\,\mu\text{A/m}$, giving a total of $37\,\text{dB}\,\mu\text{A/m}$ instead of $23\,\text{dB}\,\mu\text{A/m}$. This may make us think that better performance (longer range) is possible under Part B, but this is purely illusory.

In fact, this peak (although it is well below the limit of $42\,\mathrm{dB}\,\mu\mathrm{A/m}$) can only be reached during the maximum presence of the carrier (the logical "1"). Now, in the presence of logical "0"s – or, even worse, a long sequence of logical "0"s (needed to make the system work!), because of the maximum modulation index of 14% that must be allowed for, the carrier will only rise to 75.4% of its maximum value, that is, $28\,\mathrm{dB}\,\mu\mathrm{A/m}$ rather than $37\,\mathrm{dB}\,\mu\mathrm{A/m}$ as we had hoped. Moreover, during this time (in the presence of zeros), the energy transmitted to the transponder integrated circuit (the smart card) is proportional not to the value of the carrier but to its square, which reduces the operating range or distance accordingly (by about 58%) and which means that, while the ETSI standard is still complied with, the type A and type B operating distances are identical – subject to the important qualification that the noise immunity is much greater in 100% ASK modulation than in 10% ASK modulation.

Examples

Let us return to our initial example, based on the fact that, to conform to ISO 14 443 and operate at a distance of 10 cm, a card must operate with $H = 1.5\,\mathrm{A/m}$ at $d = 10\,\mathrm{cm}$, that is, $H_{10\,\mathrm{cm}} = 1.5\,\mathrm{A/m}$.

(a)　Let us examine a first case in which the radius r of the base station is specified, for example, $r = 10\,\mathrm{cm}$:

$$P_1 = 0.203 \times (1 + 1^2)^2$$

$$= 0.812\,\mathrm{W}$$

$$= 812\,\mathrm{mW}$$

$$a = \frac{d}{r} = 1$$

$$H_0 = H_{10\,\mathrm{cm}} \times 2.828$$

$$= 1.5 \times 2.828$$

$$= 4.242\,\mathrm{A/m}$$

(b)　Conversely, in a second hypothesis, let us assume that the amplifier can deliver only a certain power P_1, and calculate the value r of the radius of the base station antenna for (by way of example) $P_1 = 250\,\mathrm{mW}$.

This example is based on the power (approximately 23.5 dBmW at 30 Ω, in other words $P_1 = 250\,\mathrm{mW}$), which can be delivered by the well-known Philips Semiconductors MF RC 500 integrated circuit.

$$0.250 = 0.203 \left[1 + \left(\frac{0.1}{r} \right)^2 \right]^2$$

$$1.231 = \left[1 + \left(\frac{0.1}{r} \right)^2 \right]^2$$

$$0.110 = \left(\frac{0.1}{r} \right)^2$$

$$0.331 = \frac{0.1}{r}$$

therefore

$$r = 0.30\,\text{m}$$

$$a = \frac{d}{r} = 0.33$$

and

$$H_0 = H_{10\,\text{cm}} \times 1.17$$

$$= 1.5 \times 1.17 = 1.75\,\text{A/m}$$

6.4.2 The case of the ISO 15 693 vicinity standard and ISO 18 000–3

Let us make the same kind of assumption as before (subject to the same comment, of course) about this standard.

If we assume that the antenna of an ISO 15 693 base station delivers a maximum field strength equal to the authorized maximum, in other words 5 A/m, and that the same vicinity transponder again operates with a magnetic field strength of 0.15 A/m, then, for the same reasons as before, we find

$$\frac{H_0}{H_d} = \frac{5}{0.15} = 33.33 = \sqrt{(1 + a^2)^3}$$

therefore

$$a = \frac{d}{r_1} = 3.15$$

If, for example, we want to have a vicinity operating distance of $d = 60$ cm, it is necessary that $r_1 = d/a = 60/3.15 = 19$ cm, making the antenna diameter approximately 40 cm.

— The surface area of the antenna is then

$$N_1 = 1$$

$$S_1 = 1256 \times 10^{-4}\,\text{m}^2$$

(equivalent in a first approximation to a rectangular antenna of 40×30 cm).

— The radiation resistance of this antenna is then

$$Ra = 0.13(12.56 \times 10^{-2})^2$$

$$= 20.5 \times 10^{-4}$$

If we assume that $H_{0\,\text{max}} = 5$ A/m (and therefore, since the value $a = d/r_1$, we have 0.15 A/m at 60 cm with a radius r_1 of 19 cm), we can calculate the current $I_{1\,\text{max}}$ flowing in the base station antenna:

$$H_{0\,\text{max}} = \frac{N_1 I_1}{2r_1}$$

therefore

$$I_1 = \frac{H_{0 \, max} \times 2r_1}{N_1}$$

therefore

$$I_{1 \, max} = \frac{5 \times 2 \times 0.19}{1}$$
$$= 1.9 \, A \, rms$$

as well as the power radiated by it:

$$Pa_{max} = Ra \cdot I_1^2$$
$$= 20.5 \times 10^{-4} \times 1.9^2$$
$$= 7.4 \, mW$$

It should be noted that, with this radiated power, it is easy to make the spectra of the radiated electrical and/or magnetic fields conform to the ETSI or FCC limits by using balanced output stage structures (with a balun device, for example) and special technologies, such as, "figure of 8 coil antennae" for base stations, as will be demonstrated in Chapter 8 which deals with antenna technology.

ISO 15 693 in the minimum field of 0.15 A/m

To conform to the ISO 15 693 standard, the label that we have used in the previous examples (antenna measuring 4.5×4.5 cm with 7 turns) must also be capable of operating when subjected to a minimum effective magnetic field of $H_{d \, min} = 0.15$ A/m. Note that the specification of this minimum magnetic field strength does not in any way determine the distance d from the base station at which the label can operate!

As regards the label in our example, using an I·CODE circuit, we know that its threshold magnetic field is $H_{dt} = 51.2$ mA/m (already three times more sensitive than the value of 150 mA/m required by the standard), since this magnetic field causes the appearance of $V_{ic} = 2.2$ V, a value typical of the correct operation of the integrated circuit. Clearly, in the worst case, the calculations must take into account the most unfavourable value of V_{ic}, which is 3, xx V, but for now we will retain this value.

The above section still adds nothing to our knowledge of the possible operating range. We must therefore make some further assumptions.

On the principle that, for simple commercial reasons, a manufacturer of ISO 15 693 base stations will try his hardest to obtain the longest operating range, while ensuring that there is no risk of "frying" a transponder according to the standard if it happens to be pressed against the antenna, the manufacturer will set the maximum magnetic field strength to the maximum authorized by the standard, namely, $H_0 = 5$ A/m, since this will ensure that it is always possible to read the "quietest" labels in a reliable way.

Given the general magnetic field equation $H(d, r)$ with which you are now very familiar (I hope so, anyway), when we reach the value of 0.150 A/m, and therefore when the ratio H_0/H_d is as follows $5/0.15 = 33.33$, we find that $a = d/r_1$ will be 3.15 (see *Table 1.1*) and therefore the distance d will be approximately three times the radius r_1 of the base station antenna. Now you can choose the value of r_1 which will give you the distance d that you want. You want an operating distance of 60 cm? Fine, then use a radius of 20 cm. Too big? Bad luck, it's that or nothing!

This distance d will be the operating range for a label which just conforms to the standard because it operates at exactly 150 mA/m.

On the other hand, the label which we have taken as our example needs only approximately 50 mA/m to operate; in other words the ratio $H_0/H_d = 5/0.050 = 100$, where $a = d/r = 4.5$, will make it possible

— either, for the same desired operating range ($d = 60$ cm), to have a smaller base station antenna radius than that described in the previous section ($r_1 \approx 13.5$ cm);

— or, with the same antenna radius ($r_1 = 20$ cm), to have a longer operating range ($d \approx 90$ cm, that is, 50% greater than before!).

This clearly demonstrates the fuzzy nature of the term "vicinity" used in the ISO 15 693 standard. In fact, in the above example, for a circular antenna with a diameter of 40 cm (surface area 1256 cm^2), or alternatively one measuring 30 × 40 cm (giving practically the same surface area), a label whose antenna consists of seven turns in the 4.5 × 4.5 cm format used in strict conformity with ISO 15 693 (5 A/m–150 mA/m) would operate over a range of 60 cm, and the same label fitted with a more sensitive integrated circuit, for example, an I·CODE circuit, would operate over a range of 90 cm.

Now let us assume that this label is used with a base station according to ISO 15 693, and that this produces a maximum magnetic field strength of 5 A/m. In this case, let us estimate the dimensions of the base station antenna required to make this label operate at 75 cm, for example.

Given that, in this case, the ratio H_0/H_d is equal to $5/0.15 = 33.33$ this means that $a = d/r_1$ equals 3.15, in other words that, in order to have an operating range of $d = 75$ cm, we need a base station antenna with a radius $r_1 = 23.3$ cm. When $N_1 = 1$ turn in the base station, we find

$$H_0 = \frac{N_1 I_1}{2r_1}$$

$$5 = \frac{1 \times I_1}{2 \times 0.238}$$

therefore

$$I_1 = 2.38 \, \text{A}$$

The power Pa radiated by the base station will be equal to

$$Pa = Ra \cdot I_1{}^2 = \text{const}(N_1 s_1)^2 \cdot I_1{}^2$$

$$= 0.13 \times (1 \times 3.14 \times 0.238^2)^2 2.38^2$$

$$\approx 23.3 \, \text{mW}$$

Unfortunately, this application cannot be implemented easily without a special licence, since this value is rather too high to conform to the current radio frequency pollution standards.

This naturally leads us to consider how we can meet the current standards.

How to meet the current standards and optimize a vicinity application   Being aware of this latent non-conformity, let us now estimate the minimum value of the maximum operating distance, which enables us to meet the minimum specification of ISO 15 693 exactly, while remembering that we must comply strictly with the current regulations (in terms of radio frequency pollution, human exposure to non-ionizing radiation, etc.).

Constraints imposed by ETSI 300 330   As mentioned above, French legislation, in the form of the ART imposes limits on the use of SRD known as "non-specific", such as those used in RFID applications operating in the 13.56 MHz ISM band.

The limit that particularly concerns us at present is the one relating to preventing pollution of the radio frequency spectrum. ETSI 300 330 stipulates that, in specific measurement conditions (in terms of measurement methods, limits, etc.), regardless of the radiated power, the strength of the magnetic field strength at 10 m radiated by the base station must not exceed $H_{\max \text{ ETSI } 10\,\text{m}} = 42\,\text{dB}\mu\text{A/m}$ (i.e., converted to an equivalent electrical field strength, $E_{\max \text{ ETSI } 10\,\text{m}} = 47.4\,\text{mV/m}$; see the earlier discussion in this chapter) – *Figure 6.26*. Also, since it can easily be proved that, for a flat circular antenna, the relationship between the far electrical field and the radiated power is

$$E \approx \frac{7\sqrt{Pa}}{d}$$

we find

$$Pa_{\max} \approx \left(\frac{E_{\max \text{ ETSI } 10\,\text{m}} \cdot d}{7} \right)^2$$

$$= \left(\frac{47.4 \times 10^{-3} \times 10}{7} \right)^2$$

$$= 4.585\,\text{mW}$$

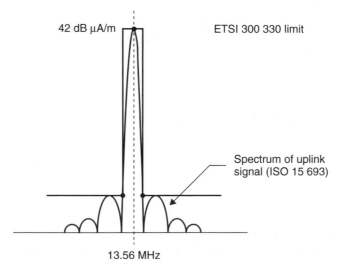

Figure 6.26   ETSI 300 330 limit and long-range applications

Constraints imposed by ISO 15693 and optimization of the system

While bearing in mind the above constraints due to ETSI 300 330, let us now continue with our example with reference to the application of ISO 15693, by returning to the equations

$$Pa = \text{const}(N_1 s_1)^2 \cdot I_1{}^2$$

and

$$H_0 = \frac{N_1 I_1}{2r_1}$$

Taking I_1 from the second equation and substituting its value in the first, we find

$$Pa = \text{const}(N_1 s_1)^2 \left(\frac{H_0 \times 2r_1}{N_1} \right)^2$$

Now let us replace s_1 with its value $(\pi r_1{}^2)$. After simplification and reduction, the latter equation takes the form

$$Pa = \text{const} \cdot \pi^2 \cdot H_0{}^2 \times 2^2 r_1{}^6$$

that is, finally

$$r_1 = \sqrt[6]{\frac{Pa}{\text{const} \times 4\pi^2 \cdot H_0{}^2}} = f(Pa, H_0)$$

Let us calculate the value of r_1 with the following parameters:

const $= 0.13$

$H_0 = 5$ A/m maximum value according to ISO 15693

$Pa_{\text{max}} = 4.585$ mW maximum value authorized by the ETSI standard,

that is,

$$r_{1\,\text{max}} = \sqrt[6]{\frac{4.585 \times 10^{-3}}{0.13 \times 4\pi \times 5^2}}$$

$$= 18.15 \,\text{cm}$$

and therefore, for the same surface area, $S_1 = 1034 \,\text{cm}^2$, an equivalent rectangular antenna of approximately 25×42 cm.

Note: I have been asked more than once why this rectangular format is preferable to a circular one. The answer is very simple: this physical configuration of the antenna fits very easily into an attaché case, and is much more easily transported for demonstrations in the field!

Moreover, since ISO 15693 defines the value of the minimum ratio H_0/H_{min} $((H_0)/(H_{\text{min}}) = 5/0.15 = 33.33)$, this immediately requires a knowledge of the parameter $a = d/r_1 = 3.15$. We are now only a small step away from finding the minimum value of the maximum distance supported by the standard, that is,

$$d = 3.15 r_1$$
$$= 3.15 \times 18.15$$
$$= 57.2 \, \text{cm} \ldots$$

a value, which finally characterizes the minimum distance associated with the term "vicinity" in ISO 15 693, assuming strict conformity with the current European legislation (according to ETSI 300 330), with a conventional antenna. In this case, we can write

$$Pa_{\max} = 4.585 \, \text{mW rms}$$
$$= Ra \cdot I_{1 \, \max}^2$$

therefore
$$4.585 \times 10^3 = [0.13 \times (1 \times 3.14 \times 0.1815^2)^2] I_{1 \, \max}^2$$

and therefore
$$I_{1 \, \max} = 1.816 \, \text{A rms}$$

The magnetic field strength in the centre of the base station antenna will then be as follows:

$$H_{0 \, \max} = \frac{N_1 I_1}{2 r_1}$$
$$= \frac{1 \times 1.816}{2 \times 0.1815}$$
$$= 5 \, \text{A/m}$$

Q.E.D.
and, for example, at $d = 3.15 r_1 = 57.2 \, \text{cm}$ the magnetic field strength will be

$$H_{57.2} = \frac{5}{33.33}$$
$$= 150 \, \text{mA/m}$$

Q.E.D.
This assumes that the transponder used is capable of operating at an induction of

$$B_{57.2} = \mu \cdot H_{57.2}$$
$$= 188.4 \, \text{nT}$$

Q.E.D.
Assuming that the transponder has a specified maximum physical surface area of $5 \times 5 \, \text{cm}$ (i.e. a maximum antenna surface area of $4.5 \times 4.5 \, \text{cm}$) and that the inductance of its antenna is $5 \, \mu\text{H}$ for the purpose of tuning the integrated circuit, and given also that the minimum typical operating voltage of the integrated circuit must be $V_{\text{ic}} = 2.2 \, \text{V}$ rms (for the I_{code} integrated circuit), we can calculate the minimum number of turns

N_2 that the transponder must have if it just conforms to the standard. To do this, we start by assigning a number to the magnetic flux:

$$\Phi_2 = L_2 s \cdot i_2$$

$$= L_2 s \cdot \frac{I_2}{Qp_2}$$

At the operating threshold of the integrated circuit (R_{ic} very high), we have shown that the current I_2 essentially flows through the integrated capacitance C_{ic} of the integrated circuit, and that we can write

$$I_2 = V_{ic}(C_{ic} \cdot \omega)$$

which gives us

$$\Phi_2 = \frac{L_2 s \cdot V_{ic}(C_{ic}\omega)}{Qp_2}$$

On the other hand,

$$\Phi_2 = B_d \cdot N_2 \cdot s_2$$

and therefore

$$B_d \cdot N_2 \cdot s_2 = \frac{L_2 s \cdot V_{ic}(C_{ic}\omega)}{Qp_2}$$

giving us, finally, the value of N_2:

$$N_2 = \frac{L_2 s \cdot V_{ic}(C_{ic}\omega)}{(B_d s_2)Qp_2}$$

$$= \frac{5 \times 10^{-6} \times 2.2 \times 23.2 \times 10^{-12} \times 85.15 \times 10^6}{188.4 \times 10^{-9} \times 20.25 \times 10^{-4} \times 28.4}$$

$$= 2.01$$

and therefore, expressed as an integer, 2 turns.

IMPORTANT NOTE

Much earlier in this book, we assumed and specified that the value of Pa_{max} was 10 mW, not 4.858 mW as stated above, since there are many handy methods for implementing base station antennae (see Chapter 8), which, while conforming to ETSI 300 330, enable us to increase the radiated power Pa to approximately 10 mW, and thus increase the operating range of a near-field system.

In this case ($Pa = 10$ mW), we would have found $r_{1\,max} = 20.7$ cm and $d = 3.15 \times r_1 = 65.2$ cm. This circular antenna with a radius of 20.7 cm is essentially equivalent to a rectangular antenna having the same surface area ($3.14 \times 0.207^2 = 1345$ cm^2) measuring approximately 30×45 cm.

ISO 15 693 in the maximum field of 5 A/m Up to this point, I have essentially examined only the behaviour of transponders operating in the presence of weak magnetic fields, in order to be sure that their remote power supply operates correctly and that communication can be assured between the base stations and transponders.

We now need to examine the other aspect of the application, namely, the use of strong fields. This is because there are many applications in which the transponders move within the space containing the magnetic field, and can therefore be subjected to high magnetic field strengths when they physically approach the base station antenna.

Smoke signals?   As I mentioned before, most base station manufacturers supply base stations which can operate over the longest possible range without ever exceeding the maximum value specified by the standard, regardless of the sensitivity of the transponders which may be present in the magnetic field.

Clearly, if strong fields are present, it is preferable for the transponder *not* to communicate with the base station by means of smoke signals! The following text is intended to help you avoid any disasters in this field that may sometimes become a "burning" question.

The ISO 15 693 vicinity standard spells out the fact that, for a label to merit the "15 693" designation, "the proximity transponder, ... must be capable of operating continuously for magnetic field strengths H in the range of 0.15 A/m to 5 A/m effective" ... and therefore it must be able to withstand them!

Let us return to our example, using a label with the physical format of $5 \times 5\,\text{cm}$ and having $N_2 = 7$ turns.

Now, assuming that the system is to operate in the presence of the strongest magnetic field, the sinusoidal current flowing in the base station antenna produces in the centre of the antenna, on its main axis, a maximum effective magnetic field strength $H_{0\,\text{max}} = 5\,\text{A/m}$, in other words, a magnetic induction equal to

$$B_{0\,\text{max}} = \mu \cdot H_{0\,\text{max}}$$
$$= 4\pi \times 10^{-7} \times 5$$
$$= 6.28\,\mu\text{T}$$

The magnetic flux captured by the antenna of the label ($4.5\,\text{cm} \times 4.5\,\text{cm} \times 7$ turns) is then equal to

$$\Phi_0 = B_0(N_2 s_2)$$
$$= 6.28 \times 10^{-6} \times (7 \times 20.25 \times 10^{-4})$$
$$= 89\,\text{nWb/m}^2$$

This collected flux, $\Phi_0 = B_0 \cdot N_2 s_2$, induces in the antenna of the transponder (having an inductance of $L_2 = 5\,\mu\text{H}$) a sinusoidal current defined by

$$\Phi_0 = B_0(N_2 s_2) = L_2 s \cdot i_2$$

and therefore

$$i_2 = \frac{\Phi_0}{L_2 s}$$
$$= \frac{89 \times 10^{-9}}{5 \times 10^{-6}}$$
$$= 17.8\,\text{mA rms}$$

... which is true!

Assuming that the global quality factor Qp_2 of the transponder is high (for a long time I have recommended an optimal value Qp_2 of 28.4, because of the value of the threshold field strength H_{dt}), then at first sight the current I_2 flowing in the input pins of the integrated circuit should be

$$I_2 = i_2 \cdot Qp_2$$
$$= 17.8 \times 28.4 \, \text{mA}$$

and therefore

$$I_2 \approx 500 \, \text{mA} \ldots \text{Help! Fire!}$$

... but this is completely wrong!

Sadly, this very simplistic argument is completely erroneous, for many reasons that will now be explained.

The function of the regulator and its results In order to be able to provide a wide choice of operating range, nearly all the transponders available on the market have internal regulation systems of one kind or another, designed to keep the voltage at the input of the integrated circuit practically constant, based on the strength of the magnetic field surrounding the transponder.

Figure 6.27 shows a typical example of these devices. The regulator shown here is of the parallel "shunt" type, but you should be aware that many others exist.

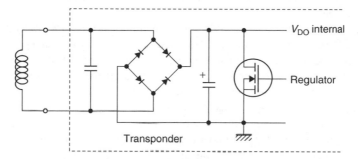

Figure 6.27 Example of shunt regulator

Let us look at the operating principle of this regulation system.

In the presence of a very weak magnetic field When the distance is large (in other words, when the magnetic field is very weak), the voltage present at the terminals of the integrated circuit is too weak to reach the threshold of activation of the regulator, and thus follows in a linear way the present value of the magnetic field (or the coupling). In this case, the integrated circuit does not yet have a sufficient internal power supply voltage to operate, and consumes no power (or very little); thus it has a very high input impedance. The transponder system then shows a very high value of Qp_2.

If the transponder is brought closer to the base station, the magnetic field strength increases and, before the regulator activation threshold is reached, the transponder integrated circuit begins to be supplied and therefore begins to operate.

In the presence of a weak magnetic field Moving still closer to the base station, the voltage at the terminals of the transponder integrated circuit reaches the specified level for correct operation, a level at which, usually, the regulation circuit has not yet begun to operate. This is the voltage (2.2 V) used throughout this book to define the maximum minimum operating range. In this mode of operation, the impedance of the transponder is known (in our case, it is 24.2 kΩ), and therefore, at this distance, the value of Qp_2 is known. In our example, its optimal value must be 28.4.

In the presence of medium and strong fields Above this threshold value, the process becomes quite different.

Regardless of the increasing strength of the magnetic field, the regulation circuit comes into action and tries (successfully!) to keep the alternating voltage at the input terminals of the integrated circuit constant. To do this, the regulation circuit progressively decreases the resistance connected in parallel with the input of the integrated circuit, thus progressively "squashing" the incident signal from the base station.

Clearly, the apparent input impedance of the integrated circuit will be modified (decreased).

We can now see that, in the far field, the structure of this impedance is principally "capacitive", since the value of R_{ic} is very high, and that it progressively becomes predominantly "resistive" because the value R_{ic} decreases markedly. The quality factor Qp_2 is therefore greatly reduced, and the initial level of 28.4 found in weak fields becomes a distant memory.

Back to the future   To make everything absolutely clear, let us return to our favoured example.

The specified maximum acceptable effective input current (in the input terminals) of the I·CODE integrated circuit is 30 mA. Note, by the way, that nobody has mentioned whether the structure of this effective current is reactive or not. It is effective, which is all that matters. Whether it flows in a capacitance or a resistance is of little importance... but this is where all the difference lies!

In fact, in the presence of a weak magnetic field (at long range), as mentioned and demonstrated above, because of the very high resistance R_{ic} (24.2 kΩ), virtually all the current (5 mA) flowing in the input terminals of the integrated circuit passes principally (4.98 mA) into the integrated capacitance (23.2 pF) and therefore a very small part (90 μA) of this current flows in the 24.2 kΩ resistance of the integrated circuit. On the other hand, in strong fields, the current flowing in the input terminal consists mainly of the current flowing in the small impedance presented by the regulator.

To quantify all the elements of the transponder in strong fields, we must start from the following two assumptions:

—   The regulator operates perfectly. It therefore maintains a constant alternating voltage at the input of the integrated circuit V_{ic} (in our case, approximately 4 V rms);

—   The manufacturer of the integrated circuit specifies the maximum effective value which the input current of the circuit must not exceed (in our case, 30 mA).

These assumptions immediately lead us to the following two conclusions:

—   On the one hand, if, regardless of the strength of the magnetic field in which the transponder is located, the voltage $V_{ic} = 4$ V is still present at the terminals of the integrated circuit, and if the input capacitance C_{ic} has not changed (and there is no reason why it should, since the applied voltage is constant), there will always be a current of $(V_{ic})/(C_{ic} \cdot \omega) = 9\,mA$ flowing in it.

— On the other hand, if we want to avoid exceeding the authorized maximum input current (30 mA effective) in the integrated circuit, the rest of the current must go somewhere else, namely, into the resistance R_{ic} provided by the regulator in this case.

This is self-evident. To be more precise, we could write div I = 0, which looks smarter, but really comes down to the same thing!

However, we should note that, in this extreme case, this remaining current (with allowance for the phase relations between the capacitive and resistive current – see the insert below), equal to approximately 28.9 mA, flows in the equivalent resistance provided by the regulation circuit.

NOTE

Example of calculation of the equivalent resistance R_{ic} presented by the regulator in the presence of a strong magnetic field (see Figure 6.28a

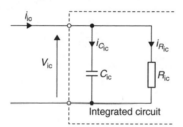

Figure 6.28a

At the input of the integrated circuit, in terms of current equations, we can write

$$i_{R_{ic}} = \frac{V_{ic}}{R_{ic}}$$

and

$$i_{C_{ic}} = V_{ic}(j \cdot C_{ic} \cdot \omega)$$

When $C_{ic} = 23.2\,pF$,

$$i_{C_{ic}} = V_{ic} \cdot j(23.2 \times 10^{-12} \times 85.15 \times 10^6)$$
$$= V_{ic} \cdot j(1.97 \times 10^{-3})$$

and therefore

$$i_{ic} = i_{R_{ic}} + i_{C_{ic}}$$
$$= V_{ic}\left[\frac{1}{R_{ic}} + j(1.97 \times 10^{-3})\right]$$

or alternatively, as an effective value:

$$I_{ic\,rms} = V_{ic}\sqrt{\left(\frac{1}{R_{ic}}\right)^2 + (1.97 \times 10^{-3})^2}$$

Assuming, for example, that

(a)   because of the regulator, V_{ic} remains constant: $V_{ic} = 4$ V rms

(b)   the maximum value of $I_{ic\,rms}$ is known: $I_{ic\,rms} = 30$ mA rms

we can deduce the value of R_{ic} from the above equation. Thus, we find that

$$\frac{I_{ic}^2}{V_{ic}^2} - (1.97 \times 10^{-3})^2 = \left(\frac{1}{R_{ic}}\right)^2$$

$$\frac{(30 \times 10^{-3})^2}{4^2} - (1.97 \times 10^{-3})^2 = \left(\frac{1}{R_{ic}}\right)^2$$

$$56.25 \times 10^{-6} - 3.9 \times 10^{-6} = \left(\frac{1}{R_{ic}}\right)^2$$

$$52.35 \times 10^{-6} = \left(\frac{1}{R_{ic}}\right)^2$$

and therefore

$$R_{ic} = \sqrt{\frac{1}{52.35 \times 10^{-6}}}$$
$$= 138.2\,\Omega$$

therefore

$$I_{R_{ic\,rms}} = \frac{V_{ic}}{R_{ic}}$$

$$= \frac{4}{138.2} = 28.9\,\text{mA rms}$$

We can again draw two main conclusions... and another one as well:

—   This current, flowing in the resistance equivalent to the regulator, produces heat in the integrated circuit as follows: $P(\text{watts}) = V_{ic} \cdot I_{R_{ic}} = 4 \times 28.9\,\text{mA} = 115.6\,\text{mW}$ (compared with approximately $200\,\mu\text{W}$ in a weak field);

—   If the input voltage is controlled – and therefore constant – and the current flowing in the regulator is known, we can determine the equivalent ohmic resistance that it represents, as follows:

$$\frac{4\,\text{V}}{28.9\,\text{mA}} = 138.2\,\Omega = R_{ic}$$

—   Finally, now that we know the ohmic resistance of the regulator, we can calculate the new value "for strong fields" of Qp_2 for the L, C, R circuit of the transponder, thus:

$$Qp_2 = \frac{Rp}{L_2 s \cdot \omega}$$

where, of course,

$$Rp = Rpc//R_{ic} = (24.2 \times 10^3)//138.2 \approx 137.2\,\Omega$$

and therefore

$$Qp_2 = \frac{137.2}{5 \times 10^{-6} \times 85.15 \times 10^6}$$

$$= \frac{137.2}{475}$$

$$= 0.288 \ ! \ldots \text{instead of the level of 28.4 we wanted.}$$

The quality factor is practically zero, and therefore, in the presence of strong fields, the global transponder circuit is totally aperiodic. In fact, with a rather low value of Qp_2 due to the low value of Rp (as a result of the low value of R_{ic}), the circuit is not tuned at all. To put some approximate but reasonably accurate numbers on this concept, let us quickly calculate the impedance $Z_{C_{ic}}$ of the integrated tuning capacitance C_{ic}. This is as follows:

$$Z_{C_{ic}} = \frac{1}{C_{ic} \cdot \omega}$$

$$= \frac{1}{23.2 \times 10^{-12} \times 85.15 \times 10^6}$$

$$= 506.17\,\Omega$$

In fact, the low impedance of R_{ic} (138.2 Ω) largely short-circuits the effect of C_{ic} with which it is connected in parallel.

As a first approximation, let us disregard the presence of C_{ic} in the system and assume that, in strong fields, the transponder is no longer tuned, but is totally aperiodic. The electrical circuit can then be summarized in the form of an LR network (*Figure 6.28*). In this case, we can write that the induced voltage V_{20} must be

$$V_{20} = \sqrt{(L \cdot \omega)^2 + Rp^2} \cdot i_2$$

Assuming the maximum $i_2 = i_{2\,\text{max}} = 30\,\text{mA}$, we obtain

$$V_{20} = \sqrt{(425.75)^2 + 138.2^2} \times 30 \times 10^{-3}$$

$$= 13.4\,\text{V}$$

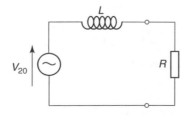

Figure 6.28b   Aperiodic strong signal transponder (LR)

that is,

$$V_{ic} = \frac{R_{ic}}{\sqrt{(L \cdot \omega)^2 + Rp^2}} V_{20}$$

$$= \frac{138.2}{447.6} V_{20}$$

$$= 0.3 \times 13$$

$$= 4.1 \, \text{V}$$

... which, as a first approximation, is reasonably accurate for a theoretical level of 4 V.

On this assumption, if the induced voltage V_{20} is 13.4 V, we can calculate H_{max}, knowing that

$$V_{20} = M_{max} \cdot \omega \cdot I_1$$

$$= \left(\frac{N_2 \cdot \Phi_{2\,max}}{I_1}\right) \omega \cdot I_1$$

$$= B_{max} \cdot N_2 s_2 \cdot \omega$$

$$= \mu \cdot H_{max} \cdot N_2 s_2 \cdot \omega = \Phi_2 \cdot \omega$$

and therefore

$$H_{max} = \frac{V_{20\,max}}{\mu \cdot N_2 s_2 \cdot \omega}$$

$$= \frac{13.4}{4 \times 3.14 \times 10^{-7} \times 7 \times 20.25 \times 10^{-4} \times 85.15 \times 10^6}$$

$$= 8.76 \, \text{A/m}$$

We could have found the same answer by another method:

$$\Phi_{2\,max} = L_{2}s \cdot i_{2\,max}$$

$$= \mu \cdot H_{max} \cdot N_2 s_2$$

and therefore

$$H_{max} = \frac{L_{2}s \cdot i_{2\,max}}{\mu \cdot N_2 s_2}$$

$$= \frac{5 \times 10^{-6} \times 30 \times 10^{-3}}{12.56 \times 10^{-7} \times 7 \times 20.25 \times 10^{-4}}$$

$$= 8.42 \, \text{A/m}$$

... which is roughly the same.

Returning to our specific initial example ($N_2 = 7$ turns) and $s_2 = (4.5 \times 4.5 \, \text{cm}^2 = 20.25 \, \text{cm}^2$), and assuming a maximum magnetic field strength H_{max} of 5 A/m according to ISO 15 693, the induced current $i_{2\,\text{max}}$ in the aperiodic circuit (because Qp_2 is very small!) is 17.8 mA.

Because of the input voltage regulation, in order to find the current flowing in the regulator, we must subtract the 5 mA, which generally flows through the capacitance. Thus, approximately 16 mA flows in the resistance R_{ic} equivalent to the controller. This means that the controller acts as a resistance of

$$R_{ic} = \frac{4\,\text{V}}{16\,\text{mA}} = 250\,\Omega$$

Figure 6.29 summarizes the operation and properties of the regulator.

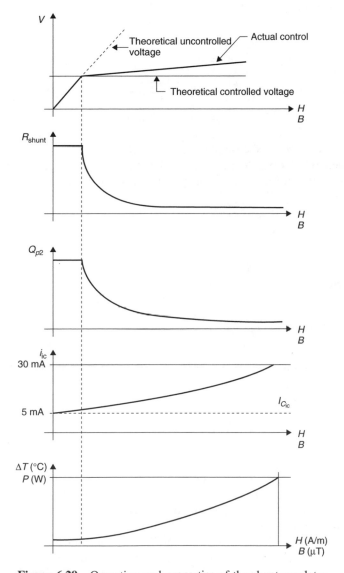

Figure 6.29 Operation and properties of the shunt regulator

Important conclusion   The discussion above clearly reveals the difficulties that have traditionally been encountered by system designers, and indicates to potential users all the complexities involved in the precise detailed specification of a system. It also clearly shows that the global electrical structure of the transponder impedance varies markedly with the distance (between the maximum and minimum distance), changing from a quasi-pure LC tuned circuit to a quasi-pure aperiodic LR circuit, and that the global value of Qp_2 of the transponder undergoes major modifications.

Purists may also note that, when the LCR arrangement and the Qp_2 of the global circuit change from an LC type with little resistive effect to an LR type with little capacitive effect, the exact value of the intrinsic resonant frequency of the transponder varies in the same way, following the precise formula for a damped tuned circuit, called the "Thomson formula".

A reminder:

$$\omega^2 = \frac{1}{L \cdot C}$$

$$\Delta = \sqrt{a^2 - \omega^2}$$

$$= \sqrt{(a - \omega)(a + \omega)}$$

where $a = R/2L$, a being the damping coefficient.

For damped oscillations to be present, the discriminant Δ must have imaginary roots, in other words $(a - \omega) \cdot (a + \omega)$ must be negative, or $(a - \omega)$ must be negative, and therefore

$$a - \omega < 0$$

$$a < \omega$$

therefore

$$R < 2\sqrt{\frac{L}{C}}$$

In the context of our example,

$L_1 = 5\,\mu\text{H}$

$C = 25\,\text{pF}$

$\Rightarrow 2\sqrt{\frac{L}{C}} = 900\,\Omega$

$R = R_{\text{ic}} = 250\,\Omega$

When the tuning changes in this way, due to the major change in the value of R_{ic}, then since the electronic threshold of operation of the regulation system is theoretically totally aperiodic, this increasing detuning between the frequency of the carrier received from the base station and the intrinsic resonant frequency of the transponder also affects the value of the real current induced in the transponder antenna. Because of the combination of all these effects, the figures given above have to be corrected slightly.

Maximum current, maximum power and temperature

Maximum current Let us start by examining the maximum current flowing in the input terminals of the integrated circuit.

For the maximum magnetic field strength of $H = 5\,\text{A/m}$ $(B = 6.8\,\mu\text{T})$ in the example considered above, we have shown that it is easy to calculate the current induced in the transponder antenna winding $(I_2 = 17.8\,\text{mA rms})$ and thus to discover whether or not we have exceeded the maximum permitted level. In our example, we are in complete conformity with the transponder specification which states a maximum value of 30 mA rms. This match with the specification means that the chip will "behave" well, since the component manufacturer will have checked that everything is correct, including the power dissipated by the integrated circuit at this point.

Power and temperature We have shown that, in the presence of strong fields, the maximum power dissipated by the integrated circuit of the transponder is of the order of

$$P = \frac{V_{\text{ic}}^{2}}{R_{\text{ic}}}$$

$$= \frac{4^{2}}{138.2}$$

$$= 115.6\,\text{mW}$$

With such a power of approximately $P = 120\,\text{mW}$ and a thermal resistance of the bare crystal of the integrated circuit $R_{\text{th}_{j-mb}}$ of the order of 50 C/W, this means, according to the thermal Ohm's law – $\Delta T = P \cdot R_{\text{th}}$ – an increase in the crystal temperature of

$$\Delta T = 120 \times 10^{-3} \times 50$$

$$= 6°\text{C}$$

Now, in real applications using $5 \times 5\,\text{cm}$ electronic labels applied to packaging boxes, when the input current is of the order of 20 mA rms (corresponding to a magnetic field strength H of approximately 5 A/m), maximum crystal temperatures of the order of 55°C are frequently found (using infrared measurement probes). This means that the temperature has risen by 30°C above the normal ambient level of 25°C.

This also means that the sum of the thermal resistances $R_{\text{th}_{\text{total}}} = R_{\text{th}_{j-amb}} = R_{\text{th}_{j-mb}} + R_{\text{th}_{mb-amb}}$ between the ambient level and the integrated circuit junction is of the order of

$$\Delta T = 30°\text{C} = 115.6 \times 10^{-3} R_{\text{th}_{\text{total}}}$$

and therefore

$$R_{\text{th}_{\text{total}}} = 250°\text{C/W}$$

that is, the thermal resistance $R_{\text{th}_{mb-amb}}$ of the packaging in this case is

$$R_{\text{th}_{mb-amb}} = R_{\text{th}_{\text{total}}} - R_{\text{th}_{j-mb}}$$

$$= 250 - 50$$

$$= 200°\text{C/W}$$

and a   hot spot of $[25°C + (200 \times 120 \times 10^{-3})] = 49°C$ develops on the packaging under the integrated circuit.

Sometimes this temperature (49°C) is enough to blacken the paper of some labels containing transponders, and/or generate wisps of blue smoke, depending on the type of packaging, if the packages are left too close to a base station antenna for too long. (When a conveyor breaks down, there will always be one package and its electronic label positioned right next to the most powerful base station antenna of the whole installation – "Murphy's Law No. 243 642")

Now that we have defined the electrical parameters of the antenna, let us go on to look at its actual technical implementation.

Part Four
Antennae and their technology

This fourth part describes the main problems associated with the methods and technology for implementing the antennae of transponders and base stations.

In fact, where the day-to-day aspects of contactless radio frequency identification are concerned, the antennae construction methods are far from being simple or established, and it is often preferable to examine this problem fully before making any decision.

The two chapters in this part of the book do not aim to solve all the problems, but rather to help every reader to find his own way through the maze of possible solutions.

Readers requiring detailed calculations for the design and construction of antennae are referred to much more specialized works, with the proviso that the impressive equations presented in many books are never 100% accurate reflections of the humble reality.

Having said this, let us go to work – there's plenty to be done!

[1] For further details, see the author's previous book on RFID.

RFID and Contactless Smart Card Applications   D. Paret
© 2005 John Wiley & Sons, Ltd

7

The Transponder Antenna and its Technology

The transponder antenna can be constructed in many different ways. The first that comes to mind is the traditional method of winding copper or aluminium wire in a circular coil, but, for reasons of cost, thickness, mechanical flexibility, and so on, many other technologies and geometrical shapes are also used in the field of contactless devices.

7.1 The Range of Technologies

Regardless of the geometrical shape – square, circular, or rectangular – which is used, antenna technology can be divided into two aspects, one being the technology used to form the coil winding and the other being the technology involved in the connection of the antenna to the integrated circuit and, if necessary, the connection of an additional tuning capacitance of the transponder.

7.1.1 Technologies for producing the coil winding

There are several possible ways of producing the winding of the antenna coil. The list of the main technological solutions can be summarized as follows:

— coil antenna made from copper or aluminium wire:
  — air-cored,
  — wound on ferrite rods to concentrate the field lines;
— etched antennae, formed:
  — either on a rigid support such as a conventional printed circuit, or on a flexible support suitable for the production of flexible labels;
— printed antennae consisting of a layer of conductive ink deposited on a rigid or flexible support.

RFID and Contactless Smart Card Applications   D. Paret
© 2005 John Wiley & Sons, Ltd

7.1.2 Connection technologies

The technology used to connect the antenna winding to the integrated circuit is an essential element in the provision of a service life appropriate to the use of the transponder. The choice of technology for constructing the winding is an integral part of the choice of technology for providing the connections to the integrated circuit. We shall mention here the methods of *bonding, flip chips* and *bumped chips* (see *Figure 7.1*).

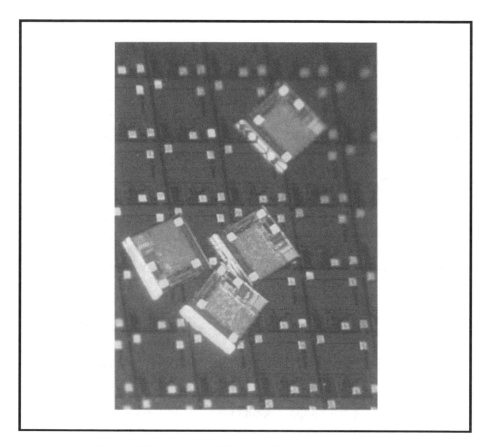

Figure 7.1 Example of "bumped" chips for transponders

7.1.3 A summary

Table 7.1, which is not claimed to be exhaustive, provides a brief summary of the principal properties of the range of technologies used to form transponder antennae.

Many companies in the market are ready and willing to solve (or help you to solve) the problems associated with the production of "inlets".

Table 7.1 Comparison of the different antenna technologies for inlets

Antenna technology	Coil	Embedded	Etched	Etched	Printed
Conductive layer	Copper	Copper	Copper	Copper	Silver-based conductive ink
Type of inter-connection	Thermo-compression bonding	Thermo-compression bonding	Soldering	Adhesive conductor	Adhesive conductor
Quality of inter-connection	Very good	Very good	Very good	Average	Average
Tear-off force (in cN)	>200	>200	>200	<60	<60
Antenna Insulation	Yes	Yes	Yes	Optional	Optional
Ease of handling during production	Average	Very good	Very good	Good	Good
Quality of conductive layers	Very good	Very good	Very good	Very good	Good
Possibility of adding a further component	No	No	Possible	Possible	Possible

7.2 The Geometrical Shapes of the Windings

Leaving aside the production technology, the most common geometrical shapes are as follows (see *Figure 7.2*):

— in the form of flat coils, wound with very thin wire:

   — circular,

   — square,

   — rectangular;

— zero thickness (except for the thickness of the printed circuit, the ink deposit, etc.):

   — circular,

   — square,

   — rectangular;

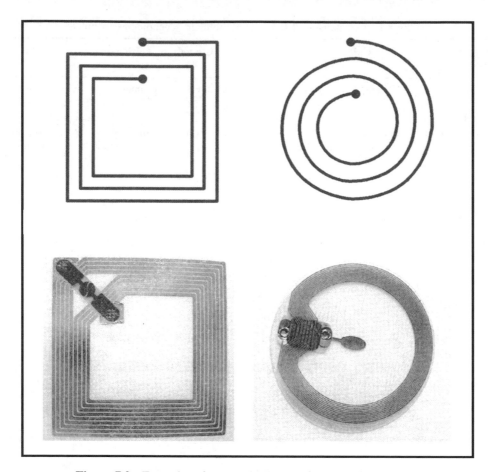

Figure 7.2 Examples of geometric shapes of transponder antennae

— spiral,

— fretted,

— Louis XV style, Louis XVI style and Christmas tree (yes, it really exists and has even been patented by the author),

— and so on.

We shall now take a quick look at most of these possible configurations.

7.2.1 The concept of inductance

As soon as a current flows through a conductor (in our case, a winding of N turns), this produces a magnetic field, a magnetic induction and consequently, through a surface, a magnetic flux Φ such that

$$\Phi = B \cdot S$$
$$= (\mu \cdot H)(N \cdot s)$$

Moreover, during this production of flux, the surface formed by the winding that has created the flux is itself immersed in its own flux. We can thus define a self-induction coefficient, called the "inductance", L, as the ratio between the flux and the current generating it, that is,

$$\Phi = L \cdot I$$

L is expressed in henrys, and therefore

$$\Phi = L \cdot I = B \cdot S$$

Note that, in passing, we can deduce from this the unit of permeability, which I have avoided mentioning for a long time, waiting for the right moment.

Thus, μ is expressed as:

$$\mu = \frac{L \cdot I}{H \cdot S} \text{ and therefore as } \frac{\text{henrys} \times \text{amperes}}{\underbrace{\text{amperes} \times \text{m}^2}_{\text{m}}} \text{ that is, in henrys/m}$$

We can use the above equations to deduce the general expression for the inductance:

$$L = \frac{(\mu \cdot H) \cdot (N \cdot s)}{I}$$

Now, regardless of the shape – circular, rectangular, flat, short, long, and so on – the magnetic field strength H is always proportional to the product $(N \cdot I)$, and therefore, for circular shapes, $H = (\text{const}(N \cdot I))/r$, an expression in which the constant *const* depends solely on the physical dimensions of the conductor. Replacing H and s with their respective values, we obtain

$$H = \frac{(\mu \cdot \text{const} \cdot N \cdot I) \cdot (N \cdot \pi r^2)}{I \cdot r}$$

and therefore, in the last analysis,

$$\boxed{L = \text{const} \cdot \mu \pi r N^2}$$

7.2.2 Inductance of a circular air-core coil

Returning to specific examples, let us consider the case of a flat air-core coil with a radius r. If the diameter D of the wire forming is very small with respect to the diameter of the coil, that is, $2r(D/r < 0.0001)$, it can be demonstrated that the global magnetic field produced by the winding is as follows:

$$H = \frac{\ln\left(\dfrac{2r}{D}\right)}{\pi r} N \cdot I$$

(not to be confused with the field strength at the centre of the antenna, which is $(N \cdot I)/2r$).

Then,

$$L = \frac{\mu_0 \cdot \ln\left(\dfrac{2r}{D}\right) \cdot N^2 \cdot s}{\pi r}$$

Replacing s with its value and multiplying above and below by r, we find

$$L = \frac{\mu_0 \cdot r \cdot \ln\left(\dfrac{2r}{D}\right) \cdot N^2 \cdot \pi r^2}{r \cdot \pi r}$$

$$\boxed{L = \mu_0 \cdot r \cdot \ln\left(\frac{2r}{D}\right) \cdot N^2 \quad \text{with } r \text{ and } D \text{ in metres and } L \text{ in henrys}}$$

or alternatively

$$L = 4\pi \times 10^{-7} r \ln\left(\frac{2r}{D}\right) N^2$$

Now let us express the value of r in cm and call this value r'. To maintain the integrity of the equation, we must multiply the second term by 10^{-2}, that is,

$$L = 4\pi \times 10^{-7} r' \times 10^{-2} \ln\left(\frac{2r}{D}\right) N^2$$

$$= 4\pi \cdot r' \cdot \ln\left(\frac{2r}{D}\right) N^2 \times 10^{-9}$$

and therefore

$$L \text{ (nH)} = 4\pi \cdot r' \cdot \ln\left(\frac{2r}{D}\right) N^2$$

where r' is in centimetres and r and D are in the same units, for example in centimetres as well. In this case, we can write

$$\boxed{L \text{ (nH)} = 4\pi \cdot r' \cdot \ln\left(\frac{2r}{D}\right) N^2 \quad \text{with } r' \text{ and } D \text{ in centimetres.}}$$

Let l' denote the length of a turn in centimetres. In the case of a circular turn, $l' = 2\pi r'$, we obtain

$$L = 2 \times 2\pi \cdot r' \cdot \ln\left(\frac{2\pi r'}{\pi D}\right) N^2$$

and therefore

$$L = 2l' \left[\ln\left(\frac{l'}{D}\right) - \ln(\pi)\right] N^2 \quad \text{where } l' \text{ and } D \text{ are in cm and } L \text{ is in nH}$$

$$= 2l' \left[\ln\left(\frac{l'}{D}\right) - 1.144\right] N^2$$

$$= 2l' \left[\ln\left(\frac{l'}{D}\right) - k\right] N^2$$

7.2.3 Simple square, circular or rectangular antenna

For the 125 kHz and 13.56 MHz frequencies, the most common conventional antennae for short and very short distances are flat air-core coils. In a second-order approximation, as shown in the preceding text, the equation governing the value of their inductances as a function of the number of turns is as follows:

$$L = 2l \left[\ln \left(\frac{l}{D} \right) - k \right] N^P$$

$$= L_0 \cdot N^P$$

where

L = total inductance of the coil (in nH);

L_0 = inductance of one turn;

l = circumference ($= 2\pi r$) or length ($2 \times (a + b)$) of one turn of the

antenna (in cm);

D = diameter of the wire or width of the conductor track if the antenna

is made in printed circuit form (in cm);

N = number of turns;

k = correction factor, whose value depends on the geometry of the antenna

= 1.04 for rectangular antennae in ISO card format

= 1.07 to 1.16 for circular antennae (normally very close to $k = \ln(\pi) = 1.144$)

= 1.47 for square antennae.

NOTE

The factor k actually represents the Napierian logarithm of a constant C whose value is $c \cdot \pi$. The coefficient c represents the correction term, allowing for the shape and specific geometry of the winding by comparing it with a circular coil having the radius r, which is large with respect to the diameter D of the wire used. The value of c therefore varies around 1 (from 0.8 to 1.5) according to the geometrical shape:

$$\ln \left(\frac{1}{CD} \right) = \ln \left(\frac{1}{D} \right) - \ln C \text{ where } \ln C = \ln(c \cdot \pi) = \ln c + \ln \pi$$

$$\ln C = \ln c + 1.144$$

Assuming that $\ln C = k$, we obtain

$$\ln \left(\frac{1}{CD} \right) = \ln \left(\frac{1}{D} \right) - k$$

Examples

— for square antennae:

$$k = 1.47 = \ln C = \ln c + 1.144$$

and therefore

$$\ln c = 0.326 \Rightarrow c = 1.385$$

— for rectangular antennae:

$$k = 1.04 = \ln C = \ln c + 1.144$$

and therefore

$$\ln c = -0.104 \Rightarrow c = 0.901$$

$p = $ an exponent whose value depends on the antenna winding technology and therefore on the relative closeness of the turns to each other.

In fact, because of the mechanical construction of an antenna winding, the turn-to-turn coupling is never exactly 100%, and therefore the value N^2 conventionally given in the mathematical equation is not accurate in reality. We must therefore redefine the value of the exponent p, which gives

$p = 1.8$ to 1.9 for antennae made from coiled wire
$p = 1.7$ to 1.85 for etched antennae
$p = 1.5$ to 1.75 for printed antennae
$\ln = $ Napierian logarithm.

NOTE

With reference to the conventional integrated circuits available on the market and the capacitances usually incorporated in them, we can say, with reasonable accuracy, that

— for the 125 kHz frequency, we generally need to produce coils having something in the region of a hundred to several hundred turns to form the transponder antenna;

— for the 13.56 MHz frequency, just a few turns are generally sufficient to form the transponder antenna.

Examples

We will take two examples of applications operating at 13.56 MHz.

Wire coil antenna   First of all, let us look at the example of a rectangular antenna measuring 7.5×4.5 cm ($l = 24$ cm $= 240$ mm), with 4 turns, made from 0.15 mm diameter copper wire ("15/100"), to be fitted on contactless cards in the standard ISO physical format, called ID1, and using MIFARE circuits operating at 13.56 MHz (therefore $L = 3.7\,\mu H$) for applications conforming to ISO 14 443. We find

$$L = 2l\left[\ln\left(\frac{l}{D}\right) - k\right]N^P$$

$$= 2 \times 24\left[\ln\left(\frac{240}{0.15}\right) - 1.04\right]4^{1.8}$$

$$= 48(7.37 - 1.04)12.125$$

$$= 3.68\,\mu H$$

Etched antenna   Now let us look at the example of the etched antenna, called the "reference antenna" for measurement, described in ISO 10 373–6 concerning the test method of the ISO 14 443 standard.

This has only one turn, and theoretically it represents the antenna L_0 (the inductance for one rectangular turn measuring 7.2×4.2 cm ($l = 22.8$ cm)).

$$L = L_0 = 2 \times 22.8 \left[\ln \left(\frac{22.8}{0.5} \right) - 1.04 \right] 1^{1.7}$$

$$= 2 \times 22.8(6.12 - 1.04) \times 1$$

$$= 232 \text{ nH}$$

NOTE

Theoretically, if the coupling between turns were 100%, then for the same physical dimensions the other values associated with L_0 should be proportional to the square of the number of turns, which would give

$$L = L_0 \cdot N^2,$$

and therefore, with

$$N = 4,$$

$$L_{\text{theoretical}} = 0.232 \, \mu\text{H} \times 4^2 = 3.7 \, \mu\text{H}$$

In fact, with allowance for the equation shown previously, we find that the real value of the 4-turn antenna is as follows:

$$L = 0.232 \times 4^{1.7} = 2.45 \, \mu\text{H}$$

Table 7.2 gives some examples of calculations.
See *Figure 7.3*.

IMPORTANT NOTE

Sometimes, in order to construct very small transponders (postage stamp format) whose dimensions are such that the antenna inductance is very low, it is necessary to use integrated circuits for transponders having a higher integrated capacitance, of the order of 90 to 100 pF. For example, the HC ("High Capacitance") I·CODE integrated circuit has a nominal capacitance of 97 pF, rather than the standard 23 pF, which, for the same tuning frequency, makes it possible to produce antennae with inductances 4.2 times less, of the order of $5/4.2 = 1.2 \, \mu\text{H}$.

Air-core circular coil (without ferrite)   These coils are easy to manufacture.
Figure 7.4a shows the parameters to be considered when estimating the inductance.

Coil on printed circuit   These coils are very easy to manufacture and they are also very inexpensive.

Circular   *Figure 7.4b* shows the parameters to be considered when estimating the inductance.

It is important to note that the tolerance on the thickness of the copper layer has a direct effect on the resistance of the winding, and therefore on the quality factor.

Table 7.2 For systems operating at 13.56 MHz

L (μH)	N	Physical size of the antenna, cm × cm	l, in cm, of one turn	Wire diameter or width/thickness of turn	p	Rs (in dc) Ω	Cp, pF
			Transponders				
For smart cards fitted with the MIFARE integrated circuit							
−coil	3.68	4	7.5 × 4.5	24	0.15 mm = Φ	1.8	6.07	4.7
	3.68	4	8 × 4.8	25.6	0.25 mm = Φ	1.8		
−etched (*)	0.232	1	7.2 × 4.2	22.8	500 μ × 35 μ	1.7	0.25	
−etched	3.65	4	7.5 × 4.5	24	150 μ × 35 μ	1.7	6.3	4.1
−etched	2.45	4	7.2 × 4.2	22.8	500 μ × 35 μ	1.7		
−printed	3.6	4	7.5 × 4.5	24	Depends on the ink	1.5		
For labels fitted with I·CODE integrated circuits							
in smart card format							
−coil	5	5	7 × 4	22	0.25 mm = Φ	1.8		
in 5 cm × 5 cm label format							
−etched	5	7	4.5 × 4.5	18	500 μ × 35 μ	1.7		
			Base stations				
For proximity base station							
−etched	1	3	8.5 × 5.4	27.8	500 to 1500 μ	1		20

(*) antenna described in ISO 10 376−6

Rectangular *Figure 7.4c* shows the parameters to be considered when estimating the inductance.

The same note applies.

Circular coil, on ferrite When this type of antenna is used for transponders, the magnetic field lines are concentrated because of the low magnetic reluctance of ferrite (type A 10 or TV 3C8, 3C85 radio ferrite, with μ_r of the order of 0.015).

This means that, in the presence of weak fields, the path of the flux can be concentrated in the ferrite – see *Figure 7.4d* – and therefore more flux can be provided inside the turns forming the antenna. This method is frequently used for antennae of transponders operating at long range (for rolls and steel trolleys for handling goods), or when the directivity of the transponder must be specific (for immobilizers).

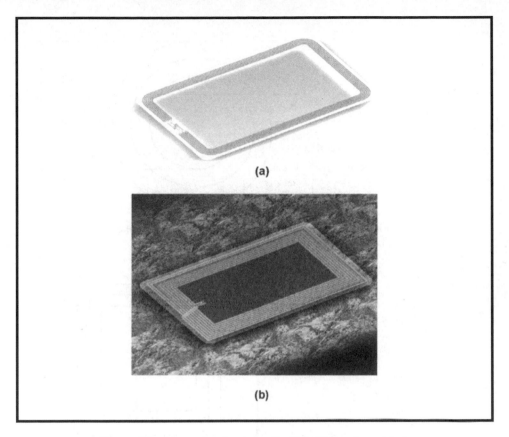

Figure 7.3 Example of antenna for ISO ID1 cards or labels

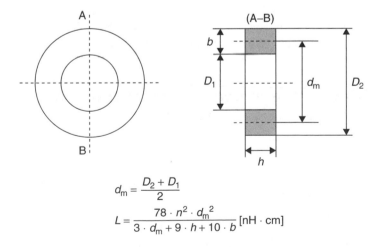

$$d_m = \frac{D_2 + D_1}{2}$$

$$L = \frac{78 \cdot n^2 \cdot d_m{}^2}{3 \cdot d_m + 9 \cdot h + 10 \cdot b} \ [\text{nH} \cdot \text{cm}]$$

Figure 7.4a Parameters for air-core coil antennae

$$d_m = \frac{D_2 + D_1}{2}$$

$$h = a \cdot n = \frac{D_2 - D_1}{2}$$

$$L = \frac{21.5 \cdot n^2 \cdot d_m}{1 + 2.72 \cdot \frac{h}{d_m}} \; [\text{nH} \cdot \text{cm}]$$

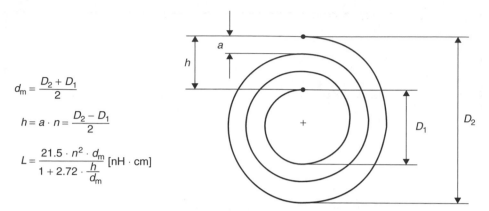

Figure 7.4b   Parameters for printed circular antennae

$$d_m = \frac{D_2 + D_1}{2}$$

$$h = a \cdot n = \frac{D_2 - D_1}{2}$$

$$L = \frac{\mu_0}{4\pi} (d_m + h) \, n^2 \cdot K_2(\xi)$$

$$\xi = \frac{h}{d_m + h}$$

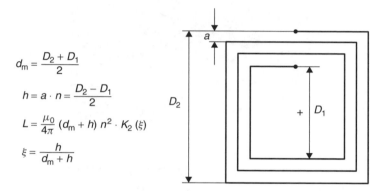

Figure 7.4c   Parameters for printed square or rectangular antennae

Figure 7.4e shows the parameters to be considered when estimating the inductance, *Figure 7.4f* presents examples of calculations and measured values.

7.2.4 Note on the tuning (accuracy and tolerance) of transponder antennae

The accuracy of the tuning frequency of the transponder is one of the fundamental parameters of an application.

In Chapter 6, I demonstrated the effect of detuning on the threshold magnetic field strength, as well as the effect of the global value of the quality factor – Q – on the expected global performance of the application. Clearly, the value of this tuning is given by the well-known formula $L \cdot C \cdot \omega^2 = 1$ and of course we must examine the performance of the two components named above, L and C (see *Figure 7.4e*).

$$\omega = \frac{1}{\sqrt{L \cdot C}} = 2\pi f$$

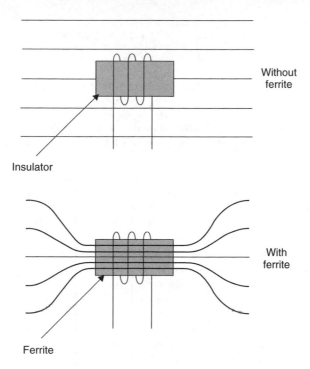

Figure 7.4d Effect of a ferrite rod on the magnetic field lines

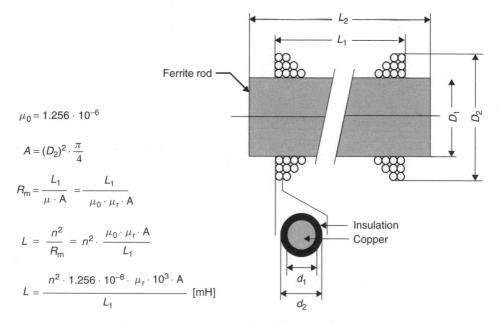

$$\mu_0 = 1.256 \cdot 10^{-6}$$

$$A = (D_2)^2 \cdot \frac{\pi}{4}$$

$$R_m = \frac{L_1}{\mu \cdot A} = \frac{L_1}{\mu_0 \cdot \mu_r \cdot A}$$

$$L = \frac{n^2}{R_m} = n^2 \cdot \frac{\mu_0 \cdot \mu_r \cdot A}{L_1}$$

$$L = \frac{n^2 \cdot 1.256 \cdot 10^{-6} \cdot \mu_r \cdot 10^3 \cdot A}{L_1} \quad [\text{mH}]$$

Figure 7.4e Parameters for circular antennae wound on ferrite rods

therefore

$$\frac{d\omega}{\omega} = \frac{df}{f} = -\left(\frac{1}{2}\frac{dL}{L} + \frac{1}{2}\frac{dC}{C}\right)$$

In the following text, we will assume that everything is operating correctly in nominal conditions.

Capacitance   The accuracy and variation of the tuning frequency is directly related to the tolerances and variations dC/C of the sum of all the capacitances involved in the

Parameter	Example 1	Example 2
D_2	21.8 mm	10 mm
D_1	18 mm	7 mm
d_m	19.9 mm	8.5 mm
b	1.9 mm	1.5 mm
h	1.35 mm	2.2 mm
n	256	488
d_{Cu}	70 μm	70 μm
l_{Cu}	14.5 m	14.2 m
L_s measured	2.295 mH	2.31 mH
R_s measured	58.9 Ω	58.5 Ω
Q measured	30.6	31.8
C_p measured	9.4 pF	7.6 pF
L calculated	2.23 mH	2.225 mH
R calculated	61 Ω	59 Ω
Q calculated	28.7	29.6

Deviation $\dfrac{(L_s - L_{calc})}{L_{calc}}$	2.9%	3.8%

Circular coil

Parameter	
D_2	2.0 mm
D_1	1.6 mm
L_2	9.4 mm
L_1	7.2 mm
d_2	47 μm
d_1	40 μm
n	532
L_s measured	2.35 mH
R_s measured	49.8 Ω
Q measured	36
C_p measured	36.2 pF
L calculated	2.28 mH
R calculated	42.6 Ω
Q calculated	42

Deviation $\dfrac{(L_s - L_{calc})}{L_{calc}}$	3%

Coil on ferrite rod
$\mu_\tau = 0.0146739$
Ferrite 3C85

Figure 7.4f   Examples of calculations and measured values

tuning; in other words, on the one hand, the parasitic capacitances due to the technology used for the antenna, the connections, and so on, and, on the other hand, the tuning capacitances properly so-called (external physical capacitors or internal capacitances of the integrated circuit).

Parasitic capacitances

Parasitic capacitance of the antenna   This capacitance arises from the mechanical spacing between the turns of the antenna coil winding, due to the presence of insulation, or, for the same reasons, between the tracks of the printed circuit. To illustrate this point, here are some orders of magnitude:

—   at 125 kHz, the capacitance is of the order of 100 pF, because of the large number of turns;

—   at 13.56 MHz, it is of the order of:
    5–7 pF for wire antennae,
    2–4 pF for etched antennae,
    2–4 pF for printed antennae.

Parasitic connection capacitance   This capacitance is due to the necessary links between the coil and the integrated circuit. The order of magnitude of this capacitance is generally several picofarads (3–5 pF).

Tuning capacitances

Fully integrated in the transponder integrated circuit   The capacitance integrated into the transponder integrated circuit is frequently fully capable of tuning the antenna in transponders operating at 13.56 MHz, since only a few tens of picofarads (approximately 25 pF) are required to tune conventional antennae, whether for ISO contactless cards or for labels measuring 5×5 cm. Indeed, for these devices, physical and cost factors rule out the provision of an additional capacitance during the final production of the component.

Another problem arises when the aim is to produce smaller labels, which naturally have smaller areas, smaller turns and lower inductances. To overcome this problem, designers are beginning to propose an integrated circuit which incorporates a capacitance having a higher nominal value (approximately 95 pF) to allow tuning without the use of any additional capacitance.

In this case, the totality of dC/C is supported by the variations and tolerances of the capacitance of the integrated circuit.

Not integrated or only partially integrated into the transponder integrated circuit
This is frequently the case with systems operating at 125 kHz. Because of the lower operating frequency, it is necessary to provide a larger tuning capacitance (of the order of 680 pF). This latter value is often disproportionate to the maximum capacitances (approximately 210 pF), which can generally be economically incorporated in integrated circuits.

To increase choice and flexibility of use, the manufacturers of integrated circuit frequently decide to provide a capacitance of the order of 210 pF for those who do not wish to add an external component, and who will therefore be satisfied with the declared tolerance of the incorporated capacitance. In this case, it will be necessary to use a high inductance with many turns, or to suggest that users provide an additional external capacitance (470 pF, for example). In this way it is possible, on the one hand, to achieve correct tuning, and, on the other hand, to distribute and weight the effects of the tolerances between the internal and external capacitances.

The latter solution has the advantage of providing an external capacitance whose tolerance and temperature coefficient can, or must, compensate for the other effects of the tolerances of the other components.

Inductance   The variations and tolerances of the inductance are principally due to

— the tolerance on the number of turns (mainly for wire-wound antennae, since this parameter cannot apply to those made in printed circuit form);

— the tolerances on the magnetic permeability of the ferrite (if there is any ferrite present).

Resistance   The variations and tolerances of the winding resistance are principally due to

— the tolerance on the number of turns (for wire-wound antennae);

— the length of the average turn;

— the tolerances on the diameter of the wire or on the thickness of the copper layer of the printed circuit.

Quality factor   The intrinsic quality factor Q of the coil is $Q = (L\omega)/R$ and therefore depends on both factors, and of course on their tolerances as well.

$$\frac{dQ}{Q} = \frac{dL}{L} - \frac{dR}{R}$$

Temperature   Most systems operating in the temperature ranges for public use (0 to 70°C) or industrial applications (−40°C to 85°C) and sometimes also in extended temperature ranges (−55°C to 125°C), where tuning must be precisely maintained in order to keep the same operating performance (in terms of range, etc.). A project designer must therefore take into consideration all the temperature dependences of the various components of the antenna.

Final Note

While on this subject, be careful about using wire-wound antennae that have to be overmoulded with plastic material. In general, the expansion and contraction coefficients of the metal wires of the winding are quite different from those of the plastic material, leading to mechanical stresses and possible breakage of the internal connections between the antenna wires and the integrated circuit, either at t_0 or over a period of time.

8

The Base Station Antenna and Its Technology

It goes without saying that the base station antenna has to be wound; but here again, there are many different methods of forming it.

Of course, the first method that comes to mind is the traditional one of winding copper or aluminium wire in a circular coil, but, for reasons of functionality (short/long range, for example), cost, dimensions, thickness, and so on, many other technologies and geometrical shapes are also used in the field of contactless devices.

The list of the main technological solutions can be summarized as follows:

— antennae wound with copper or aluminium wire:

— with air cores;

— on ferrite cores;

— antennae etched on a support of the printed circuit type;

— printed antennae, produced, for example, on thin film or paper supports.

Regardless of the construction technology used, the geometrical shapes are frequently as follows:

— in the form of flat (and thin) coils:

— circular;

— square;

— rectangular;

— zero thickness (except for the thickness of the printed circuit, the ink deposit, etc.):

— circular,

— square,

— rectangular;

RFID and Contactless Smart Card Applications   D. Paret
© 2005 John Wiley & Sons, Ltd

—   spiral,

—   fretted,

—   and so on;

—   three-dimensional:

—   XYZ antenna;

—   and so on.

We shall now take a quick look at most of these possible configurations.

Figure 8.1a summarizes the electrical and physical layout of the base station antenna.

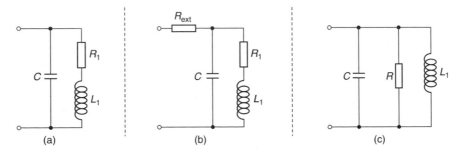

Figure 8.1   Electrical circuit of the base station antenna

It comprises:

—   the pure value of the inductance, L_1, of the antenna;

—   the value of the series resistance, R_1 (including, if necessary, its value at high-frequency (HF), allowing for the skin effect);

—   the capacitance intrinsic to the physical construction of the antenna C (capacitance distributed over the whole length of the turns, etc.), connected in parallel with everything else.

To assign some orders of magnitude, let us assume that the values of L, C and R are in the region of

	At 13.56 MHz	At 125 kHz
L	300 to 1000 nH	350 to 500 μH
C	qq pF to qq tens of pF	qq nF to qq tens of nF
R	0.x to several ohms	qq ohms to tens of ohms

This electrical circuit leads us to estimate the value of $Q_{intrins}$ (in no-load conditions) for the physical construction of the antenna. Order of magnitude of $Q_{mean} = (L_{mean} \cdot \omega)/(R_{mean})$, that is, with $L = 800$ nH, $C = 27$ pF, $R = 0, 5\,\Omega$, which, for a frequency of 13.56 MHz, gives $Q \approx 136$.

Since this value is too high and not generally suitable for the application ($Q_{\text{max useful}}$), it is often necessary to connect an external physical resistance – R_{ext} – (see *Figure 8.1b*), or to wind the antenna with a thinner wire, in order to be able to adjust the value of the global Q to meet the application requirements, with allowance for the many other parameters. Note that the power dissipation capacity of this external resistance must be specified correctly, since all the current flowing in the antenna passes through it.

After a few series/parallel transformation operations which are standard for a given frequency (see the appendix on series/parallel and parallel/series conversions), we obtain the equivalent circuit shown in *Figure 8.1c*.

In systems operating at high frequency (13.56 MHz for example), to improve power efficiency, the impedances are generally matched between the generator (the antenna driving circuit) and the load (the antenna). Unfortunately, the impedance represented by the last equivalent circuit of the antenna that I have shown you is not matched (in terms of impedance, and therefore also in terms of power) to that of the optimal generator represented by the output circuit of the base station (which is often less than 50 Ω). We must therefore provide an impedance matching circuit, for example, by using a capacitance bridge such as that shown in *Figure 8.2*.

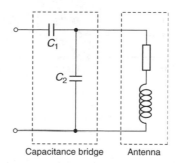

Figure 8.2 Capacitance bridge for impedance matching

Figure 8.3 shows a standard example of a system used with the MF RC 500 circuit for MIFARE applications.

8.1 Shape, Size and Technology of the Base Station Antenna

The shape and size of the base station antenna are frequently determined by the type of application envisaged and the maximum acceptable physical size of the system and base station (e.g. conventional antennae, portable smart card readers, supermarket price scanners, supermarket gates, etc.), rather than by the purely theoretical aspects that would enable us to produce an optimal solution. The design constraints are therefore numerous (volume, weight, mechanical, electrical, magnetic and climatic environment, cost, etc.).

8.1.1 Single antenna

This is the most widely used solution for short- and medium-range applications, because it is generally the easiest to implement... but it is not necessarily child's play! The

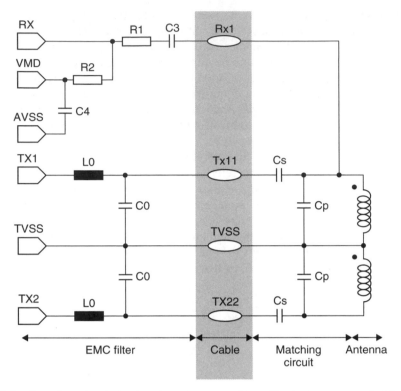

Figure 8.3 Complete circuit (tuning, impedance matching, filtering) recommended by Philips SC for MIFARE applications

following text provides an outline of the pitfalls awaiting the designers of single antenna systems.

Simple square, circular or rectangular flat antenna As described in the previous chapter, the most widely used conventional antenna for short and very short ranges is the flat coil type. Its inductance, as a function of the number of turns and the shape, is given by the equation

$$L = 2l \left[\ln \left(\frac{l}{D} \right) - k \right] N^P$$

$$= L_0 \cdot N^P$$

To avoid tedious repetition, I suggest that you refer to the preceding chapter for details of definitions, units, and comments on the parameters of this equation, which are equally applicable to base station antennae.

Figure 8.4 shows the shape of the resulting field lines.

NOTES

(1) At the 13.56 MHz frequency, one or a few turns are generally sufficient for the base station antenna.

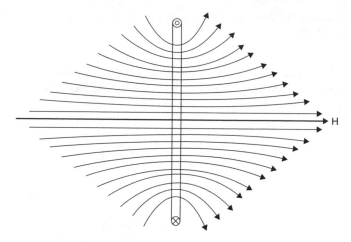

Figure 8.4 Magnetic field lines produced by a flat circular coil

(2) Square and rectangular antennae can be considered simply as variants of circular antennae (with a few tricky angles). In the equation relating the number of turns and the inductance, only the value of k is changed ($k = 1.47$).

(3) A special remark should be made here about the practicalities of constructing antennae with sharp corners, such as square or rectangular antennae. Instead of winding them with standard wire, it is often preferable to use a flat (ribbon) cable of the type conventionally used in information technology applications. *Figure 8.5* shows a simple way of forming

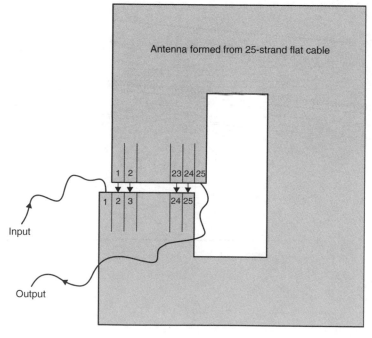

Figure 8.5 A simple way of forming a flat rectangular antenna, using 25-strand flat cable

a very flat square antenna with 25 turns, which can be fitted easily in the thickness of a door, for example.

Figure of 8 coil antenna   This winding method (see *Figure 8.6a*) has the advantage of considerably reducing – by about 15 to 20 dB – the far radiated electrical and magnetic fields (see *Figure 8.6b*).

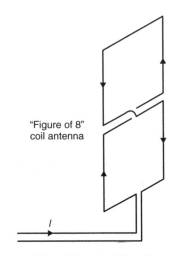

Figure 8.6a   "Figure of 8" coil antenna

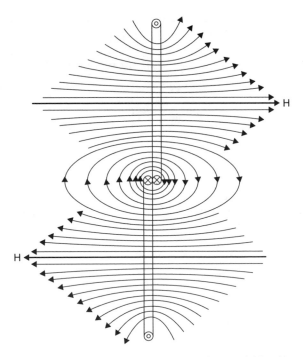

Figure 8.6b   Magnetic fields produced by "figure of 8" coil antenna

This is because the two fields produced by the two "half-coils" wound in the form of a figure of 8 are in phase opposition at each instant, so that the resulting far magnetic field strength is reduced more rapidly in the principal axis of the antenna than in the case of the simple coil mentioned above, without attenuating the strength of the near field used for the application.

An important consequence   One of the important consequences of the use of this figure of 8 coil method for forming base station antenna coils is the fact that, for the same permitted maximum level of radiation (according to the FCC, ETSI, etc.), it is possible to increase the current flowing in the base station antenna and thus provide a stronger near magnetic field, and therefore to operate over a longer working range.

Now that you know the secret, the next time you walk through an exit gate of a large store you will certainly notice that the central horizontal bar of the gate is not just a stiffening component, but is there mainly to provide a figure of 8 shape for the antenna!

Compensated, balanced, screened and shielded antennae   Regardless of whether the antenna is single or multiple, in industrial applications the user is frequently obliged to

— protect the immediate environment from perturbation created by the antenna;

— protect the antenna from undesired effects such as the near presence of metallic or magnetic bodies, or hand capacitance, and so on.

We shall go on to examine the solutions relating to these two cases, but it is important to know that, essentially, this is always a matter of investigating the paths and possible deformations of the magnetic field lines produced by the base station antenna, since, as you should remember, only the base station antenna radiates, and the transponder does not radiate at all.

Protecting the environment from perturbation created by the antenna
It is generally necessary to ensure that the electromagnetic waves emitted by the base station antenna do not perturb its immediate environment. There are a number of ways of achieving this, and these will now be described.

Compensated antenna   In most applications, the base station antenna driving stage delivers a signal that is asymmetrical with respect to earth; in other words, it has a "hot" point (the active signal) and a "cold" point connected to earth, as shown in *Figure 8.7*. Unfortunately, this configuration causes currents to flow between the various parts of the antenna and earth because of capacitive coupling.

In order to compensate for these parasitic perturbing currents, the system must be balanced as far as possible in terms of its capacity of not radiating. For this purpose, the system shown in *Figure 8.8a* is constructed; here we can see a new antenna strand that is not connected to anything, but which, because of its particular magnetic coupling to the initial active antenna, develops a pd that is equal but of opposite sign, thus providing the desired compensation.

Figures 8.8b and *8.8c* show how this is done for an industrial antenna.

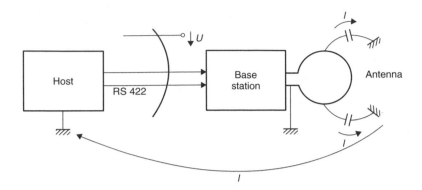

Figure 8.7   Standard antenna configuration

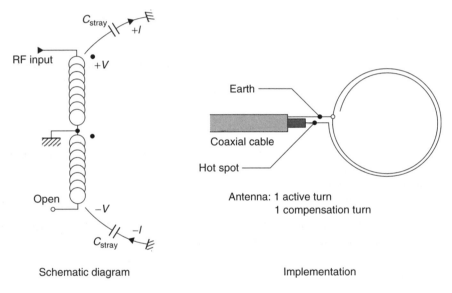

Schematic diagram   Implementation

Figure 8.8a   Antenna with compensation turn

Balanced antenna and balun system   For systems that have to deliver considerable power (base stations for long-range systems), a more thorough balancing of the circuits and currents mentioned above becomes an absolute necessity.

To facilitate the transition from the unbalanced state, due to the electrical topology of the base station output stage, to the balanced state required in the antenna driver, we will now examine the well-known "balun" system.

Figure 8.9 shows an outline of this.

As shown in Chapter 3, this circuit, based on the principle of a mid-point transformer, enables us to create two voltages in phase opposition with respect to earth, thus cancelling the undesirable effects of the flow of the currents mentioned previously.

This arrangement is often used when the antenna impedance also has to be matched to that of the control circuit, which is generally 50 Ω in 13.56 MHz applications.

Rectangular compensated antenna, 115 × 75 mm

Antenna	External components
$L = 330$ nH	$C_1 = 47$ // 3.3 pF
$C_s = 20.2$ pF	$C_2 = 270$ // 68 pF
$R_s = 0.25\ \Omega$	$R_{ext} = 0.5\ \Omega$

Layer 1

Layer 2

Figure 8.8b   Example 1

Protecting the antenna from the electrical, metallic or magnetic environment

Let us assume that the antenna is located in an electronic, metallic or magnetic environment. To prevent these environments from perturbing the action of the antenna, we must consider the use of screened and/or shielded antennae.

Antenna screened by an earth plane   In this case, a screen is inserted between the antenna and the other electronic or electromagnetic elements. *Figure 8.10* shows how

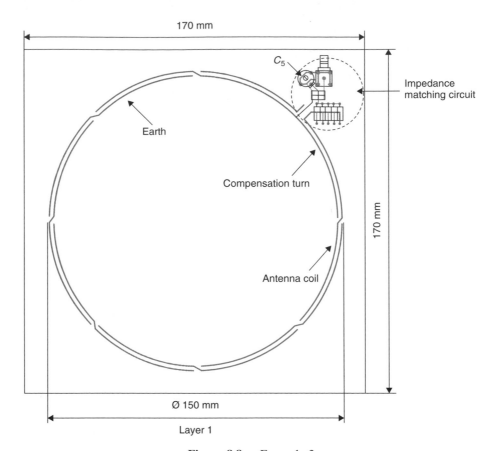

Figure 8.8c   Example 2

this is done. The top of the antenna is driven by the hot point of the output stage, and its lower part is connected to earth. The shape of the screen does not cover the whole surface of the antenna, and is intended to minimize the radiation of the electrical field, while only having a very small effect on the magnetic field required for the application.

The screen is also connected to earth, but the portion of the circuit that it forms must never be closed on itself to form a short-circuited loop (turn). Enormous earth planes are therefore to be avoided. Note that the presence of the screen also modifies the inductance of the antenna.

Antenna shielded by magnetic ferrites

Electrical environment: Let us take the example of a contactless ticket validator installed in a bus or underground train.

The electronic part of the validator is often physically located directly behind the position of the base station antenna. To ensure that the integrated logic circuits are not perturbed by the radiated radio frequency waves, it is preferable to shield them against the back radiation of the antenna. For this purpose, pieces of ferrite or plastoferrite are generally fitted; because of their low magnetic reluctance, they act as preferred paths (or even short circuits) for the magnetic field lines, keeping them as short as possible and thus preventing troublesome perturbations.

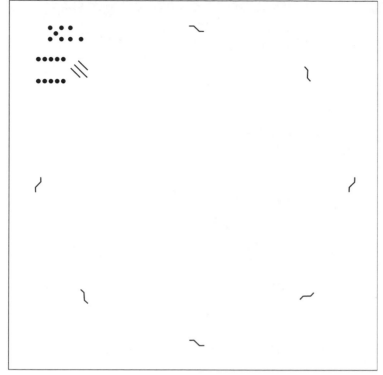

Layer 2

Figure 8.8c (*continued*)

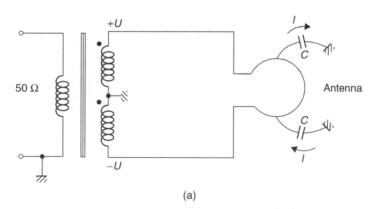

(a)

Figure 8.9a Balancing with balun transformer

That settles the problem of backward lines. Now let us look at the forward lines.
Ferrite pieces fitted in this way give the system another property. Not only do they
act as a screen for the back radiation, but, because of the deliberate guiding of the
magnetic field lines behind the antenna, some of these lines are partially projected

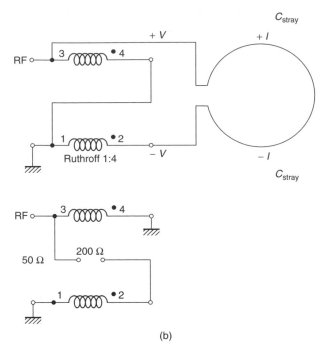

Figure 8.9b   Balancing with balun autotransformer

in front of the antenna by a "mirror effect", thus reinforcing the forward field lines, and enabling the read range to be increased, while helping to provide more precise directivity and enhanced reproducibility. *Figure 8.11* gives a clear picture of the effects obtained by this method.

Metallic and magnetic environment: The field produced by the antenna will be more or less perturbed, according to the materials forming its metallic or magnetic environment. It is therefore preferable to take all possible measures to avoid this. For this purpose, it is simply necessary to fit an element having low magnetic reluctance, such as a magnetic shield or a piece of ferrite, between the antenna and the metallic parts.

Figure 8.12 shows an example of such a device designed for a vehicle immobilizer system, in which the quality and performance of the base station antenna must be made independent of the type of material (steel, iron, brass, zamac, etc.) used in the manufacture of the steering column and its lock. Also, as mentioned before, using a ferrite ring provides better guidance of the field lines and helps to increase the directivity and operating range (see above).

FINAL NOTE

I would also recommend that a metal plate be fitted under the ferrite and then connected to earth. By doing this, you will make the antenna totally independent of its physical position, regardless of whether or not it is placed in a metallic environment. You should also be very careful to connect the earth of the metal plate to the correct HF earth point of your system... otherwise you run the risk of inventing a whole new theory of antennae!

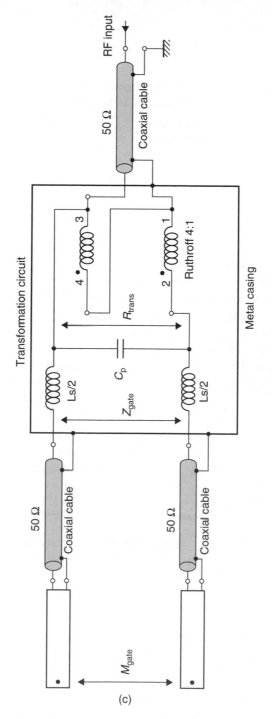

Figure 8.9c   Example of balancing with balun autotransformer

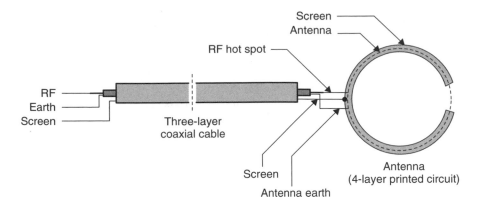

Figure 8.10a   Antenna screened by an earth plane

Lastly, and obviously, if a ferrite plate or metal screens are placed in the system to provide protection, their presence and final physical positions must be borne in mind when the antenna is tuned, and the tuning capacitances must be adjusted accordingly.

Electrical insulation of the antenna and cancellation of "hand capacitance"   The antenna coils for potential users frequently have to be electrically insulated. Clearly, the conventional techniques of enamelled wires or varnish deposited on printed circuits can be used, but sometimes, in order to provide enhanced electrical insulation at the "Class II" level, the insulation provided previously has to be improved by fitting an additional plastic sheath. The problem so far is simple enough, but it becomes more complicated when we consider the following section.

Hand capacitance and capacitive or inductive detuning   So now we have a perfectly insulated antenna that a user can touch physically without any risk of electrocution. Sadly, this is not enough, since the presence of a human body (a hand, etc.) contributes to the modification of the tuning of the tuned antenna circuit by a capacitive or inductive "hand effect". We must therefore find a suitable solution to overcome these new problems; such a solution is described below.

Antenna in aluminium tube   On the principle that it is always preferable not to electrocute users when they touch a base station antenna (you lose customers that way!), and in order to avoid "hand capacitance" problems which may disturb the tuning of the antenna, the antenna is very often completely insulated electrically from all public areas.

For this purpose, when the winding itself has been electrically insulated (with enamel, varnish, plastic sheathing, etc.) as mentioned above, the whole base station antenna coil is usually placed inside a casing, which is an electrically conductive casing but has no, or very little, magnetic effect. Aluminium is a good example. This casing is then connected to earth (see *Figure 8.13*). Thus, with the aluminium tube completely connected to earth, nobody will be electrocuted, and the hand capacitance which might have detuned the base station antenna will be eliminated.

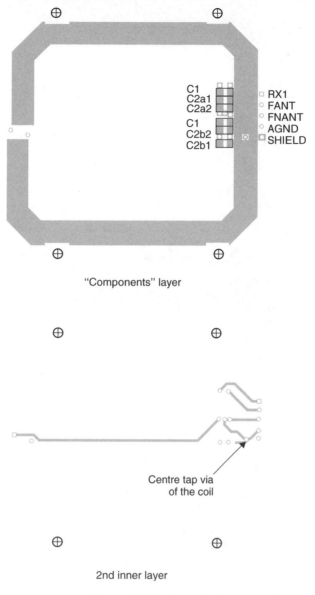

Figure 8.10b Example 1

Clearly, with this design a supplementary parasitic capacitance with respect to the electrical earth is added to the base station antenna, and this must be allowed for when determining the tuning of the base station antenna.

The main problem that arises is generally a mechanical one, as the various turns making up the base station antenna winding have to be fitted into the aluminium tube. This is relatively easy, or not too complicated, when the antenna has only a

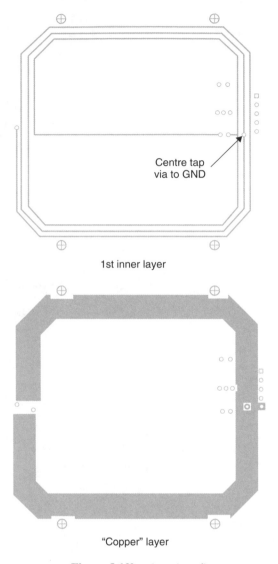

Figure 8.10b (*continued*)

few turns (1 to 5), but is often problematic when the antenna has many turns. In this case we use, not tubes, but two half-shells that are joined together electrically and mechanically afterwards.

8.1.2 Multiple antennae

These antennae are normally used for long-range applications, since they can cover large volumes (from several tens of dm^3 to several m^3) in which more than one transponder may be present simultaneously (and therefore collision management is provided).

Diameter 15 cm, 3 layers

Antenna	Components
$L = 460$ nH	$C_1 = 39 \,/\!/\, 3.3$ pF
$C_s = 38.7$ pF	$C_2 = 180 \,/\!/\, 15$ pF
$R_s = 0.53 \; \Omega$	$R_{ext} = 0.5 \; \Omega$

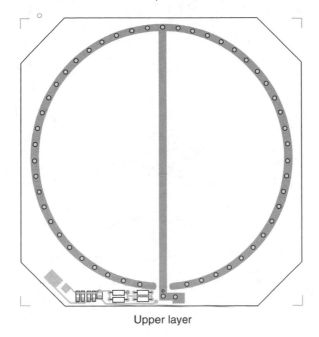

Upper layer

Figure 8.10c   Example 2

In these applications, where the base stations are concerned, we often find two classes of antennae, called "synchronized" and "multiplexed". This difference is not usually enough to determine the design of the antennae. There is another distinction to be drawn. There is the question of whether the transponders moving in the magnetic field with respect to the base station antennae have known or fixed positions, or are presented in batches, in a random way, in any direction (in the latter case, we speak of "batch reading" – as explained below). These two cases are often not as clear-cut as one would like.

Synchronized antennae   These types of multiple antennae, usually supplied in series or in parallel (and therefore having the same current or branched identical currents flowing through them), operate simultaneously, and their magnetic fluxes can be in phase or in phase opposition. To illustrate the "in phase" and "in opposition" solutions, an example of an application is shown below, relating to gate devices controlling access by persons wearing access badges on their chests (*Figure 8.14*).

In phase: Figure 8.14a shows the resultant of the magnetic field lines when these are produced by two coils that are in phase electrically and magnetically.

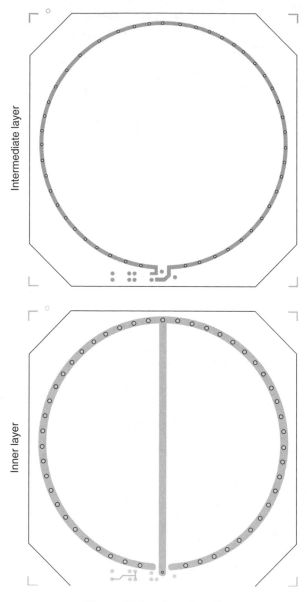

Figure 8.10c (*continued*)

This figure shows that the read range of the transponder is not as great as might be hoped, owing to the parallel orientation of the field lines with respect to the transponder plane in a large part of the space in which the magnetic field is developed.

In opposition: Assuming that exactly the same current flows in the coils (which are in series electrically) but that the direction of flow of the current is reversed in one of the coils, then the coils will be in flux opposition. We can see in *Figure 8.14b* that, for the same application (same base station, same current, same power), a much longer operating range is obtained than in the preceding case, because of the much

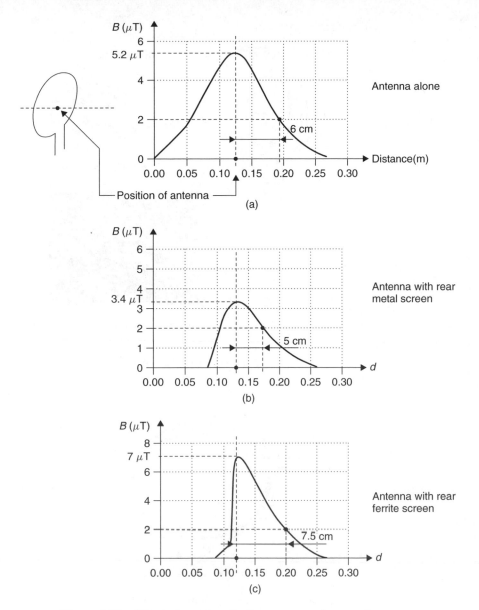

Figure 8.11 Effect of metal or ferrite screens on the operating range

larger region of space in which the field lines are orthogonal or almost orthogonal to the plane of the transponder antenna. The only magnetic "black hole" is located at the geometrical centre of the system, at the point where the magnetic fields of the two coils cancel each other out. Note that this point has no effect on the application, because there will have been plenty of time to identify the transponder beforehand.

These examples are intended to make you aware that the first solution that comes to mind is not necessarily the best!

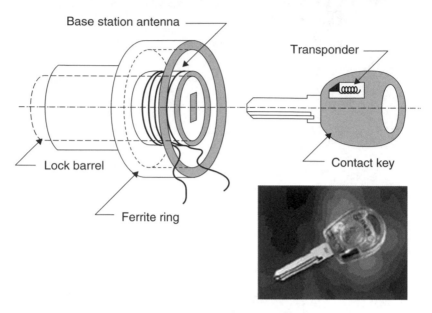

Figure 8.12   Immobilizer fitted to a steering column lock

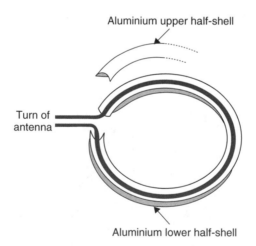

Figure 8.13   Antenna in aluminium shell

Non-synchronized or multiplexed antennae   In this configuration, in order to create numerous magnetic fields, a number of antenna $(2, 3, \ldots, 5, 10)$ are supplied by an appropriate time multiplexing method; in other words, they are supplied sequentially in time, making it possible to reduce considerably the number of base stations to be used, but also tending to lengthen the recognition times of the transponders. Note that this can also weaken the magnetic field produced, depending on the distances between the antennae. This is because, owing to the law of reciprocity, every transmitting antenna is also a receiving antenna, and when it is not radiating it absorbs energy produced by another antenna.

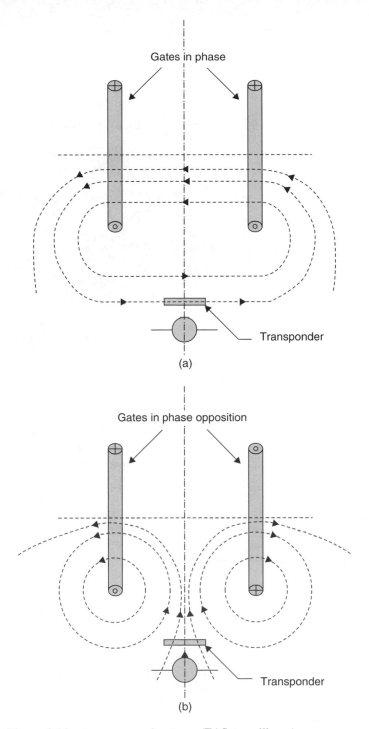

Figure 8.14   Access control antenna (EAS surveillance): access gates

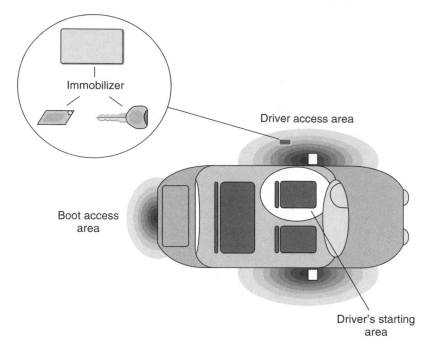

Figure 8.15   Transponders used in "hands-free" function for vehicle access

A classic example is that of "passive keyless entry" systems for motor vehicles. Since the designer does not know which side of the vehicle the driver and/or passengers will approach from, the antennae must be positioned in the front and rear, right and left doors, and in the boot (see *Figure 8.15*), and, for reasons of cost, the number of base stations inside the vehicle must be minimized.

Moreover, in this case, we must prevent the double, triple or multiple reading of the same transponder by one or other of the numerous antennae during their corresponding time slots. This is facilitated by the fact that each transponder has a unique serial number. In general, the chosen collision management principle and the application software will be responsible for preventing double reading.

Antenna for batch reading   For "batch" reading, it is necessary to be able to communicate with transponders regardless of their physical positions and orientations in the magnetic field.

To achieve this, the magnetic field must never become parallel to the plane of the transponder antenna. We must therefore think in three dimensions concerning the base station and transponder system, according to the specific nature of the application.

Let us look at three very different and very familiar cases.

Supermarket gate applications   The problem is simple. The transponder must be simple and inexpensive. The gate can be complex, because there are far fewer gates than transponders. The solution lies in the construction of base station antennae (generally "figure of 8" wound) in two and/or three dimensions (see *Figure 8.16*).

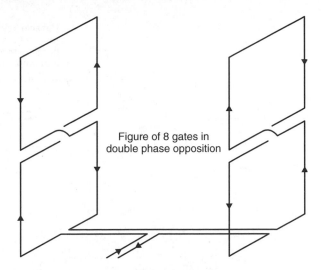

Figure 8.16   "Double-8" gates in double phase opposition

Motor vehicle applications   If we carefully examine the "passive keyless entry" function, we will soon realize that, because of simple mechanical factors, it is often difficult to provide numerous antennae in a vehicle, and especially to overcome the problems related to the three-dimensional aspect. Consequently, this problem is resolved in the transponder, by means of a three-dimensional antenna, called "cubic" or "3D", which can cut the flux in three dimensions.

Figure 8.17 is a schematic diagram of this implementation.

Industrial applications   In the case of object monitoring, it is frequently necessary to read transponders "in batches", for example, to discover the internal contents of several boxes on a single pallet.

In this case, several antennae must be positioned orthogonally with respect to each other, making the system independent of the random positions of the transponders located in the space covered by the magnetic field.

In this case also, it is difficult to supply all the antennae simultaneously, and the usual procedure, as before, is to use time multiplexing of the fields of 2, 3, 4, 5, 6, ... antennae (see *Figure 8.18*).

This technique is not (too) troublesome, because in this type of application there is usually enough time to perform this operation. The magnetic field is applied to one antenna after another, and the numerous transponders that can be supplied by the antenna in question are detected after collision management, or by a counting process. The base station software does everything else necessary for collision management and/or counting the possible double readings.

One last word on this subject. Some ingenious solutions have been developed (and often patented) in order to reduce costs. *Figure 8.19* shows an example of a design for a single near-flat antenna in which the turns of the antenna winding are positioned in such a way as to provide a "virtually" three-dimensional magnetic field, which is often largely sufficient to meet the requirements of many applications.

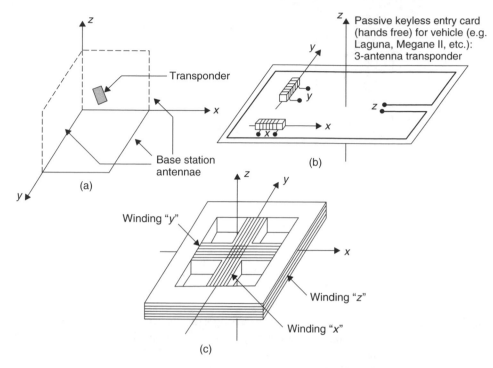

Figure 8.17   3D antenna for hands-free badges for vehicle access

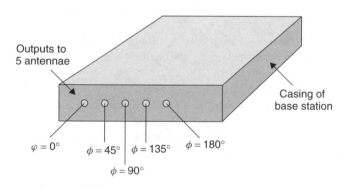

Figure 8.18   Base station with multiple phase shifted outputs

The rotating field method   Sometimes the time multiplexing described above is rather difficult to implement in specific situations. This method often requires antenna switching based on relays or MOSFET transistors in which high currents flow, with associated complex electronic control circuits.

To avoid this, we can imagine a device such as that shown in *Figure 8.20*, enabling this multiplexing to be carried out in an apparently "automatic" way without the help of any other system.

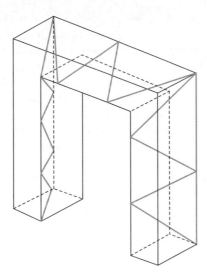

Figure 8.19 Example of skilful implementation of a "virtual 3D" antenna

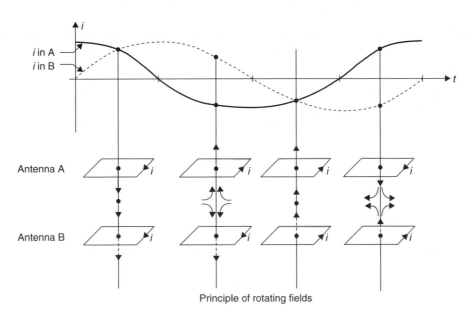

Figure 8.20 Principle of rotating field antennae and applications to gates

As shown in the figure, the two antennae are driven simultaneously by two currents, permanently phase-shifted by 90° with reference to each other. These two currents (from different sources, or with one derived from the other) are therefore present simultaneously and are therefore not time multiplexed in phase.

As shown in the figure, the magnetic fields are in phase, in phase opposition, or have opposite values, at different instants during the period of the sine wave, and this creates "rotating" magnetic field lines in space as a function of the time, with the strengths and directions of these lines varying constantly.

Because of this characteristic of instantaneous variation of the magnetic field created in this way, the whole of the space can be scanned in different directions and at different strengths, and a transponder can be identified regardless of its instantaneous position and orientation in the volume covered by the whole field (this is useful in cases where the transponders are placed "in batches", necessitating "batch reading").

Figure 8.20 shows a solution based on two antennae in a parallel arrangement, as in gates, but is also possible (if the user still has doubts) to provide a third antenna, placed orthogonally with respect to the other two, and to adjust the relative phases of the currents flowing in the different antennae... which themselves may also be wound in figure of 8 configurations.

So you are free to devise the craziest configurations and the most fantastic patents but be warned, there are already numerous patents filed in this area!

To conclude this chapter, I shall briefly review some specific applications and contactless systems.

Passive and active antennae   Here are two new special terms to define.

Generally, in contactless technology, the base stations are highly integrated (see *Figure 8.21a*); in other words, the microcontroller used for managing the communication protocol, collision management, data encryption, and so on, and the application itself is located very close to the analogue power (transmission) and reception part, and is also close to the antenna (which may give rise to problems of mutual interference). In this case, the distance between the antenna and the power stage is small.

Remote passive antenna   Sometimes, for simple practical reasons, the antenna winding has to be physically located at a certain distance from the rest of the base station electronics (see *Figure 8.21b*). In this case, we speak of a "remote passive antenna".

We may make the following remarks on this type of arrangement:

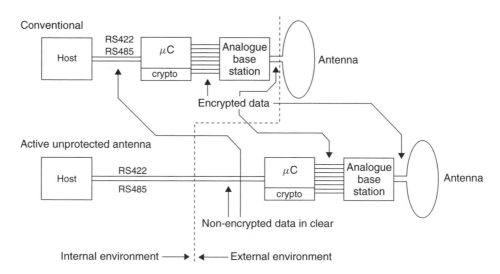

Figure 8.21   Passive, active, remote passive and protected antennae

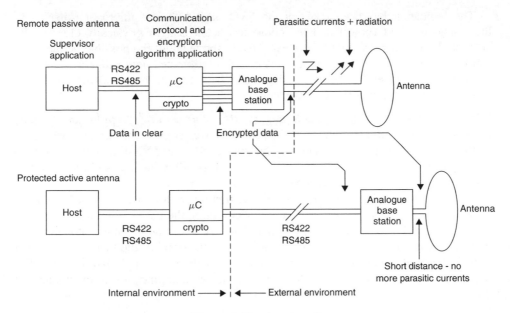

Figure 8.21   (*continued*)

— During the whole of the uplink between the base station and the transponder, a signal of considerable power is present in the connecting wires leading to the antenna, and if precautions are not taken by using twisted pairs or coaxial cable, these wires radiate and may perturb the immediate environment;

— In the downlink, in addition to the powerful signal present in the antenna connecting wires, when very weak signals from the transponder (due to the load modulation of the transponder) are received, these wires may also be subject to parasitic phenomena, possibly causing perturbation and errors in reception. The same applies, although in a less critical way, to the use of twisted pairs.

— Finally, if the signals sent between the base station and transponder have to be encrypted, you should note that they remain encrypted in the connecting wires, and that these wires do not require special security protection.

Active antenna, non-protected   Now let us imagine the layout shown in *Figure 8.21c*.

The separation has been made at the level of the host, and the "microcontroller + analogue part (power and reception)" section has been brought physically close to the antenna. In this case, the first two remarks made above about the possible perturbation can be disregarded. This configuration is known as a *"non-protected active antenna"* because the data travelling along the connecting wires between the host and the "microcontroller + analogue part" system are not encrypted and must be protected by other means.

Protected   active   antenna   Finally, let us imagine the layout shown in *Figure 8.21d*.

The separation has been made at the level of the "host + microcontroller", and the "analogue" part (power and reception) has been brought physically close to the antenna. In this case, the first two remarks made above about the possible perturbation can be disregarded. This configuration is called a *"protected active antenna"* because the data travelling along the connecting wires between the host and the "protected active antenna" continue to be encrypted.

Wake-up systems using a low-frequency (LF) carrier and high-frequency (HF) communication   Some systems use the remote power supply and the passive aspect of the uplink (from the base station to the transponder) in "passive" transponders to trigger other devices or functions of "active" transponders having additional built-in batteries (known as "battery-assisted systems").

In these systems (e.g. electronic identification numbers for sporting events, runners, triathlons, etc., and passive keyless entry devices, etc.), the base station transmits low-frequency or radio-frequency radiation modulated by a communication protocol, which may or may not be security protected, to wake up some or all of the functions provided in the integrated circuit. In this kind of system, an HF signal (433 MHz for example) is used for the communication of the return signal (or sometimes for all the uplink and downlink communications).

Part Five

The electronics involved

RFID and Contactless Smart Card Applications   D. Paret
© 2005 John Wiley & Sons, Ltd

9

Electronic Systems of the Base Station

Here is another epic chapter in the story of contactless technology!

Who knows how many questions have been asked about this area? I would need at least two more books to reply to all of them. Perhaps one day I will be brave enough to set up a web site dedicated to the (MFAQ) (Most Frequently Asked Questions) to meet all the enquiries about this very high-profile subject!

In the meantime, this chapter will attempt to indicate the common traps and pitfalls awaiting novice designers in this field.

The electronic system of a base station normally comprises two major elements (*Figure 9.1*):

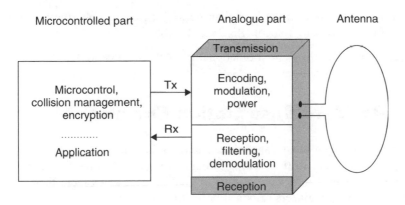

Figure 9.1   General synoptic diagram of the electronic system of a base station

The analogue part:

This is often the bugbear of systems designers. It includes:

— for transmission:

— the bit coding,

RFID and Contactless Smart Card Applications   D. Paret
© 2005 John Wiley & Sons, Ltd

— the carrier modulation,

— the power stage;

— for reception:

  — the reception of the signal produced by the load modulation of the transponder,
  — the signal processing (amplification, filtering, etc.),

  — the demodulation of the signal and the validation of the bits.

The microcontrolled part:

Generally, this part does not pose too many problems... except for the usual presence of various bugs that seem to be an unavoidable feature of information technology and programming!
  It includes:

— management of the communication protocol (transmission and reception);

— and, if necessary,

  — collision management,

  — management of the encryption algorithm,

  — display on a local LCD screen,

  — the connection to the host (if necessary), for example, the management of the connection protocol (RS 232, RS 485, etc.);

— and, quite often, the desired functionality of the application!

9.1 Standard Base Station Circuits

In the first place, there is no standard base station circuitry!
  In the second place, all the existing circuits are adapted to the requirements of the applications for which they were designed. This is what gives them their charm, and infuriates electronic engineers responsible for their design.
  Seriously, you should be aware that circuits for proximity applications have absolutely nothing in common with those for vicinity applications. The same applies to systems for managing collisions in proximity and vicinity applications. Thus, there is no magic formula, but plenty of experience (and white hairs).
  Some of you may observe (quite rightly) that base station circuits can often (or always) be found in good handbooks issued by component manufacturers. True, but these are circuits for applications "for children"! What I mean is that they are for very simple applications, for very short to short ranges (less than 5 cm or so).

These circuits usually consist of the following elements (*Figure 9.2*):

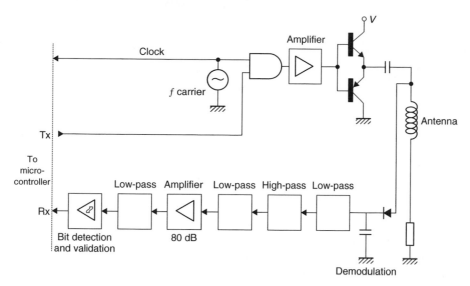

Figure 9.2   Schematic diagram of the analogue part of a base station

—   for transmission:

    —   a simple or bridge amplifier to drive the antenna circuit;

—   for reception:

    —   an amplification chain constructed from conventional operational amplifiers,

    —   band-pass filters,

    —   low-pass filters,

    —   an envelope demodulator (which is why the zero lines are present) and, in a few cases, an amplitude and phase demodulator,

    —   bit detection,

    —   bit validation;

—   for transmission and reception:

    —   a microcontroller for processing the communication between the base station and transponder, and vice versa, and rarely for collision management, encryption, connection to the host, display and other gadgets.

In short, a thoroughly conventional set-up.

Figure 9.3 (on the following double spread) shows one of these standard, traditional, academically approved base station circuits.

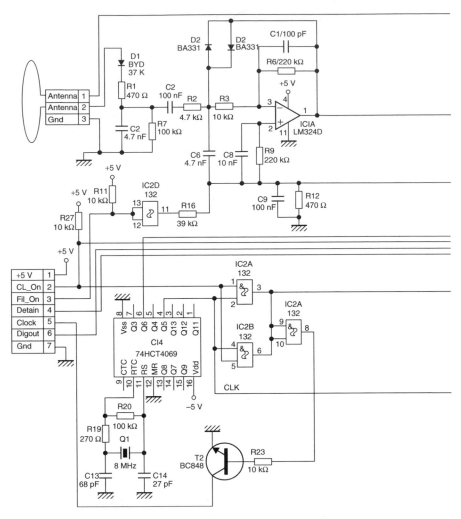

Figure 9.3 Example of a base station circuit (from a Philips SC document)

9.2 Advanced Base Station Circuits

Finally, something serious... although these base station circuits are not usually published, because they depend on major and lengthy development processes, contain many special technical arrangements and are, in principle, "privately owned". Moreover, these technical arrangements are generally protected (very well) by numerous patents, and these base stations are sold to users in the form of finished products, or in the form of operating licences.

These circuits are more specifically designed for "vicinity" and "long-range" systems.

Here we can have some more fun. Without giving away state secrets, I shall clarify some specific points, corresponding to particular examples of application, in the following paragraphs.

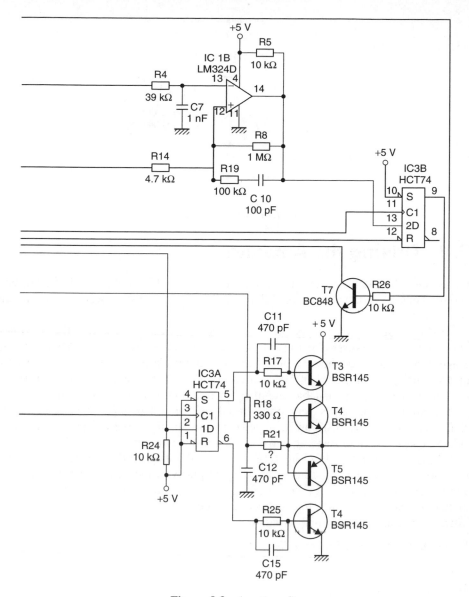

Figure 9.3 (*continued*)

9.3 The Output Stage Transmission Part

Throughout this book, I have indicated how the output stages of conventional base stations can be constructed. The only additional comment I want to make now is about long-range systems, in which output stages operating in Class C must be used to optimize the energy efficiency and minimize the power dissipated in the output transistors.

Figure 9.4a shows a standard example of the principle of such a stage.

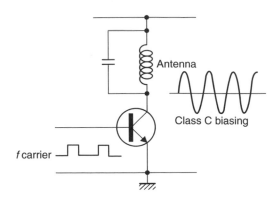

Figure 9.4a   Class C output stage

9.4   Driving the Antenna

I have already mentioned this point several times in this book, describing "direct", "non-balanced" and "balanced or differential" drive systems, as well as "balun" circuit systems.

At this stage, I shall simply point out once again that, in Europe, the maximum permitted radiated power (EIRP) is of the order of 5 to 10 mW at 13.56 MHz for long-range applications. As shown before, we must deliver a power in watts of the order of 4 W, dissipated in a load of 50 Ω, in order to reach this value; but it is not such a simple matter to design a stable amplifier for this power level at 13.56 MHz.

The amplifier design is very often based on the use of s-parameters and Smith charts, to provide the best power, impedance and stability criterion matching, and to avoid excessively high standing wave ratios and the consequent risk of bit errors.

On the other hand, the FCC and ETSI limits on the shape of the radiated spectrum do not allow much leeway in terms of the spectral purity of the emitted carrier frequency and the effect of the type of modulating signal and modulation mode on the amplitude of the sidebands. It is therefore necessary to take particular care with the design of filters for cleaning the emitted radiation.

Power matching   See *Figure 9.4b*.

9.4.1   Reception

Before investigating a few details of the processing of the received signals, let us examine some of the electrical parameters.

9.4.2   Return voltage induced in the base station antenna

As we have already explained several times, after it has sent its interrogation commands, the base station switches to listening mode for the responses from the transponder. For this purpose, the base station constantly sends a pure carrier frequency and

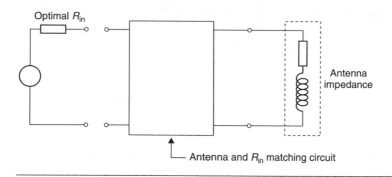

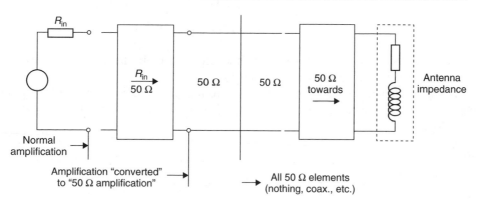

Figure 9.4b   Power matching of the various elements in the chain

waits until the transponder signals its presence and responds by means of a special modulation representing the variation of its load.

In terms of physics, due to the reciprocity of the mutual induction phenomenon between the primary and secondary, the voltage induced in the winding of the base station antenna is given by the formula, which is a reciprocal of that which gives the voltage induced by the base station in the transponder antenna (see Chapter 3), that is,

$$\Delta V_1 \text{ at the base station antenna terminals} = \Delta V_2 \cdot M \cdot \omega I_2$$

or alternatively

$$\Delta V_1 = \Delta V_2 \cdot Q_1 \cdot k \sqrt{\frac{L_1}{L_2}}$$

Clearly, this voltage decreases as the distance between the transponder and base station increases, since the coupling coefficient and the mutual inductance become lower.

For the present, to provide an idea of some of the orders of magnitude, these voltages are generally of the order of

— several hundred millivolts for proximity distances (5 to 10 cm) (coupling coefficient k of the order of a few percent);

— a few hundred microvolts for vicinity systems (50 to 70 cm) and long-range systems, of the order of a metre (k of the order of a few tenths of 1%).

An example with

$L_1 = 0.5\,\mu\text{H}$

$L_2 = 5\,\mu\text{H}$

$Q_1 = Q_2 = 30$

$k = 1\%$ if $\Delta V_2 = 5\,\text{V pp}$ then $\Delta V_1 = 500\,\text{mV pp}$

$k = 0.3\%$ if $\Delta V_2 = 5\,\text{V pp}$ then $\Delta V_1 = 150\,\text{mV pp}$

$k = 0.01\%$ if $\Delta V_2 = 5\,\text{V pp}$ then $\Delta V_1 = 5\,\text{mV pp}$

Clearly, the electronic part of the base station will have to be able to process these signals.

A comment (to be treated with caution):

When the transponder is located very close to the base station antenna, the coupling coefficient k can reach approximately 20 to 30%. In the above conditions, we would have found

$k = 20\%$ if $\Delta V_2 = 5\,\text{V pp}$ then $\Delta V_1 = 10\,\text{V pp}$

$k = 30\%$ if $\Delta V_2 = 5\,\text{V pp}$ then $\Delta V_1 = 15\,\text{V pp}$

As mentioned before, this voltage ΔV_1 is superimposed on the voltage already present at the terminals of the base station antenna (order of magnitude: 60 to 300 V pp). The resultant signal on the antenna is therefore modulated by this return signal and thus has a modulation index of the order of several percent. This signal is present at the terminals of the base station antenna and is therefore also radiated, with its specific spectrum, and its own sidebands that must, of course, be within the FCC and ETSI limits.

9.4.3 And then...

Now that we have outlined the major physical (and philosophical) question of the order of magnitude of the induced voltage recovered at the base station, we must deal with a host of equally important problems:

— how do we amplify this tiny voltage (hundreds of millivolts for short-range and proximity applications, or hundreds of microvolts for long-range applications), which is superimposed on the several tens (or hundreds) of volts already present at the antenna during the interrogation process;

— what is the value of this voltage by comparison with the interfering "noise signals" such as:

    — industrial noise picked up by the antenna (e.g. radiation from PCs, etc.),

— the intrinsic noise signal of the base station amplification chain;

— making a global estimate of the signal to noise ratio of the whole system.

We must not overlook the problems relating to

— the specific properties and qualities of the signals present, where multiple transponders are present in the magnetic field, in which case we must investigate

  — the synchronous collision management:
  at short distances,
  at long distances only, in the presence of a weak signal,
  where short and long distances are present simultaneously, in other words,
  in the presence of weak collisions;

  — the case of collision management by time slot methods:
  the possible strategies for short distances, for long distances, and for short
  and long distances simultaneously (weak signal, field strength information).

9.4.4 The amplification, demodulation and detection chain

In order to use the digital data sent by the transponder, we need to

— amplify and filter the signal;

— demodulate it;

— detect it;

— and finally, validate the value of the bit.

Before going into more detail on each of these points, I must stress to the reader that, because of the possibility of collisions between signals from different transponders, all these subjects are closely interrelated, and it is impossible to describe the receiving function of the base station in a linear way, as in the case of an ordinary receiving chain.

If you have followed the explanation from the beginning of this book, you will no doubt have noticed several times that the signal resulting from the load modulation of the transponder is very small (of the order of tens to hundreds of millivolts, or several hundreds of microvolts for vicinity or long-distance applications) and is superimposed simultaneously on several hundred volts present on the same antenna.

If we assume that the system is positioned in an ideal way in a pollution-free environment (with no parasitic signals), the ratio between the useful signal to be extracted and the transmitted signal is commonly of the order of -80 to $-110\,$dB.

In other words, we have to find a microscopic signal (a "weak signal") among enormously large signals. Clearly, this is no simple matter ... unless ...

The problem looks completely different when a high voltage (hundreds of volts) is applied to the base station antenna. The solution described previously cannot be used to recover the signal produced by the load modulation of the transponder. Other solutions must therefore be considered.

Weak signal detection   We must start by returning to the early discussion of the exact type of bit coding responsible for the load modulation used by the transponder (see also the author's previous book on RFID) because, once again, the choice of the bit coding for the downlink (from the transponder to the base station) is fundamental.

Different types of bit coding for this link are available on the market. They can be divided into two major classes, namely, those consisting of bit coding without a subcarrier (simple Manchester, etc.) and those whose bit coding requires the presence of a subcarrier (Manchester coded subcarrier, BPSK, etc.).

These two bit coding principles require two different approaches to weak signal detection and processing.

Let us start with the simpler one, for a change!

Presence of encoded bit "without" subcarrier   As mentioned several times, depending on whether the application is short, medium or long range, the voltage across the terminals of the base station antenna can be, accordingly, of the order of 30, 40, 60, 100, 120, ... or even 600 V peak-to-peak, which is rather high. We should note in passing that the same voltage is also present at the terminals of the capacitance and that it is the peak (the "hot" point of the antenna or capacitance) that is considered to be connected to the demodulator input. This causes problems, for the following simple reasons:

—   in transmission, the voltage is modulated by the base station;

—   in reception, the load modulation of the transponder causes a very slight variation in the voltage present on the antenna coil of the base station.

This means that, when the signals from the transponder are received, a pd of a few hundreds of millivolts is superimposed on the X volts present at the terminals of the coil, and in this case there is no possibility of filtering the subcarrier present in the signal ... because there is none! We must therefore deal with the whole of the signal present on the base station antenna.

This leads on to the major problem of specifying the input ratings and the sensitivity of the demodulator to be used.

If the whole of the signal present at the terminals of the base station antenna coil is taken, the input stage must be able to withstand the whole of this voltage (which is rarely the case); otherwise, we must provide a divider bridge in order to reduce the size of this large voltage, to make it compatible with the ratings of the input stage – but of course the presence of this component will similarly reduce the very small signal from the transponder (see *Figure 9.5*).

This dilemma (maximum permitted voltage versus received signal) again raises the problem of the threshold of sensitivity of the demodulator, particularly for long-range systems in which the voltage on the antenna during transmission can be of the order of 100 V d.c. (often more!), while the useful signal received from the transponder is of the order of 200 μV d.c., giving a simple ratio of $(100)/(200 \times 10^{-6}) \Rightarrow 20 \log 5 + 20 \log 10^5 \approx 114$ dB!

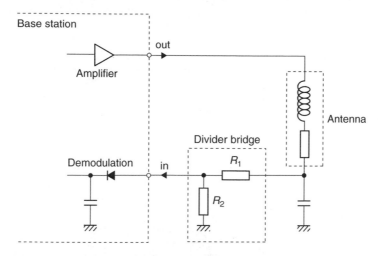

Figure 9.5   Input of the demodulation – resistance divider bridge chain

In simple cases, in other words when the voltage applied to the base station antenna is low (a few tens of volts), that is, generally when short-range systems are used, if we do not want to provide a divider bridge in parallel with the voltage on the antenna coil (or with the capacitance), we must physically connect a resistance in series with the output of the output stage of the base station, to deliberately create a voltage drop during transmission without affecting the depth of modulation present during reception on the antenna. This resistance also enables the output circuit to be protected from short circuits, as mentioned in Chapter 3.

Presence of encoded bit "with" subcarrier   If we use the load modulation technique whose bit coding includes a subcarrier frequency (e.g. Manchester encoded with subcarrier, or BPSK, as shown in *Figure 9.6a*) for the downlink, the signal can be extracted (almost) without difficulty from the noise because of the presence of many transitions, but this choice is not entirely free of penalties, as you will discover later on.

Regardless of whether the intrinsic tuning frequency of the transponder is tuned to the frequency of the carrier sent from the base station, during the load modulation of the transponder the energy spectrum of the signal present at the terminals of the base station antenna contains, in addition to the carrier frequency, two balanced sidebands located one on each side of the carrier, due to the presence of the subcarrier of the received signal (see *Figure 9.6b*).

Note: this independence is due to the fact that what are known as "synchronized" transponders (99% of the market) set their internal clocks to the signal sent from the base station, and that the frequency of the subcarrier signal is obtained by division of the incident signal.

We have just mentioned the energy content of the subcarrier. In theory, the best way of recovering the signal from the transponder is to recover the maximum of energy, in other words to recover, if possible, all the energy contained in the two sidebands by using two special frequency-calibrated filters, one at $(fc + fsc)$ and the other at $(fc - fsc)$. Clearly, this greatly complicates the receiving and filtering section, and consequently the whole electronic system of the base station.

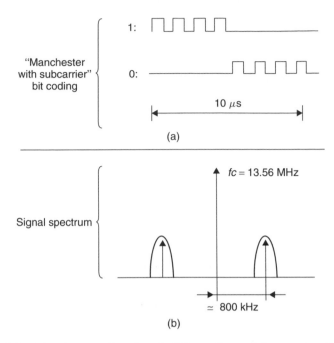

Figure 9.6 "Manchester with subcarrier" bit coding and the associated spectrum

If we only want to recover the subcarrier content, therefore, it is easier to detect the small signal resulting from the load modulation by filtering only the signals contained in these two sidebands and amplifying these, without being overwhelmed by the relatively enormous carrier signal.

Because of the cost of the filters, it is sometimes suggested that only one of the two sidebands should be processed, thus foregoing the additional energy contained in the other one... although this may make a small difference to the operating range obtained. Of course, everyone will tell you that you should choose, and use, the sideband that has more energy.

Is this the higher one? Or the lower one? A good question!

If you wish to make savings, you will need to establish which of the two sub-bands to choose, by examining the system very closely, asking yourself which band has the highest energy, bearing in mind the structural unbalance of the frequency response of the tuning of the RLC circuit (and therefore of the factor Q) of the base station antenna, and the detuning, whether deliberate or not, of the transponder in applications where transponders are arranged in stacks. We all have our problems, don't we?

Figure 9.7 summarizes a few well-known examples.

Clearly, this last section only applies to transponders allowing load modulation with a subcarrier.

A (preliminary) conclusion In short, as you will have realized, it thus becomes technically feasible to recover these weak signals in a "clean" environment.

Now we must consider industrial environments, with all their associated parasitic phenomena.

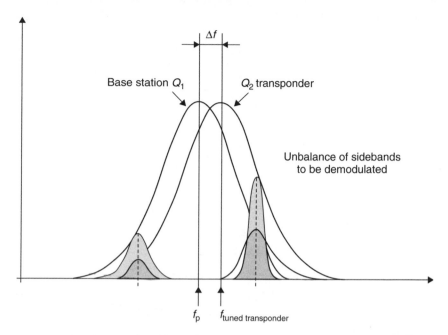

Figure 9.7   Effect of base station/transponder detuning on the signal sidebands

It will be your task to provide filters of all kinds – passive, active, analogue, digital, with Digital Signal Processing (DSP), and so on – together with preventive measures against the ambient noise level, and so on, in order to extract the very small signal from the industrial noise and to be able to say, "at last, I've recovered a signal!"

But is this signal correct? Can we check it? A large question, which we must respond to in general terms, using a special process, such as sampling the signal level several times during its period and then carrying out complementary signal processing with a majority logic method.

At this point, you will be proud that you have finally managed to communicate with just one transponder at a long distance from the base station of a system in an industrial environment. Without knowing it, you have now established a "weak detection" system and you can now say, "at last, I've recovered a valid signal!"

The block diagram in *Figure 9.8* provides a summary.

Sadly, this is not the end of your problems.

Weak collision   In fact, when working on long-range systems, given the volume covered by the base station antenna radiation, we soon find that transponders tend to move in "flocks" and it is quite common to find many of them in a single (magnetic) field, which, sooner or later, will give rise to problems of collisions in communications between the base station and the transponders.

In my previous book on RFID, I provided a detailed analysis of the general problem of deterministic, probabilistic and other collision management systems; however, there is a gap between theory and practice that may, if we are not careful, become a yawning chasm. To avoid this trap, we will now look at the tricky case that is commonly known as "the weak collisions problem".

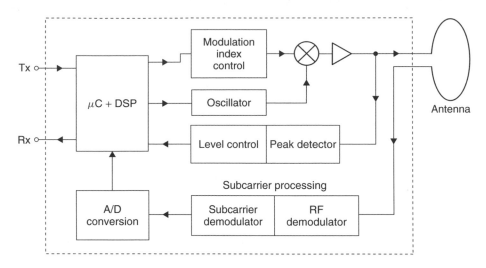

Figure 9.8   General schematic diagram of an advanced base station

In this type of application, there are five common situations:

An isolated transponder located near the base station antenna   No problem: the signal is strong, there is no collision, everything is fine. This is the classic, academic situation, described in all the textbooks.

An isolated transponder located far from the base station antenna   This is the rather more difficult case that was described above in terms of "weak detection". The signal is weak, and affected to some extent by industrial noise, but there is no collision. I have shown you how to deal with this case. It is not so "classic", but at least it is described in this book!

Two transponders located near the base station antenna   The signals generated by the load modulation are powerful. We now have to consider the standard problem of managing the possible collisions of messages. You can now choose among the standard collision management systems – deterministic, probabilistic, and so on. Theoretically, everything will be fine (*Figure 9.9a*). And in reality too! This was explained in detail in my earlier book!

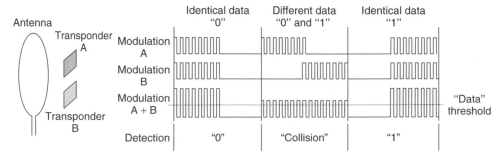

Figure 9.9a   Signals received in the case where the two transponders are near the base station

Two transponders located far from the base station antenna   This gets more complicated. We begin to encounter some serious difficulties. This is because the two transponders are far from the base station antenna and the received signals, produced by load modulation, are weak and at the limit of the ambient industrial noise. The base station amplification chain does as much as it can; it amplifies the signal to the maximum of its capacity. This means that I have omitted something important from my previous remarks – I deliberately avoid saying that every well-designed amplification chain has an automatic gain control (AGC) for automatically adjusting the amplifier gain according to the level of the incident signal.

In the preceding cases, there was no problem; the signals were powerful and "clean", the AGC reduced the gain to avoid saturating the amplifiers, and everything was fine. In the present case, the chain amplifies to the maximum level not only all the small useful signals, but also the surrounding noise (*Figure 9.9b*).

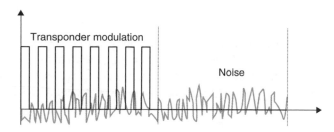

Figure 9.9b   Noise occurring when the transponder is far from the base station

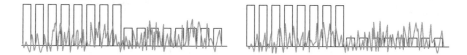

Figure 9.9c   Signals received when the two transponders are far from the base station

Let us look at the case shown in *Figure 9.9c*, where there is a collision of the synchronous type, enabling us to define a fast deterministic collision management method. Since the noise signal is practically at the same level as the useful signal, we have the problem of knowing whether there is a collision.

A standard method for dealing with this problem consists in measuring the prevailing noise level, quantifying it, weighting it, and so on, in the absence of the transponder response, in order to discover how to manage the effect of the noise in the presence of collisions.

One transponder located near the base station antenna and one located far from it   We now come into a very difficult area.

Let us look at the case where one transponder is located near the base station antenna and the other is remote from it. The signals due to the load modulation are markedly different; the first is powerful, but the second is weak, at the limit of the ambient industrial noise. An example of this very common situation is shown in *Figure 9.10*.

Because the AGC of the base station amplification chain is not very intelligent, it cannot tell the difference between a strong signal and a weak signal when they are both

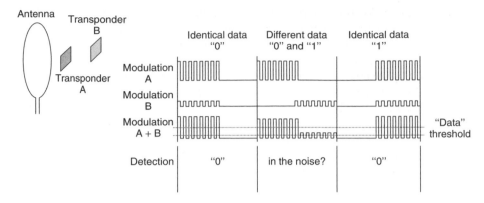

Figure 9.10 Signals received in the case where one transponder is near the base station and the other is far from it

present. Consequently, believing that it is just dealing with a strong signal, it reduces the gain to a minimum to avoid saturating the amplification chain.

Conclusion: The signal from the nearer transponder is set to a reasonable level and the signal from the far transponder and the noise signal become even weaker than before.

We must now examine two subsidiary cases.

Deterministic collision management mode   *Figure 9.11* shows what happens when we examine a synchronous collision, enabling us to define a fast deterministic collision management method.

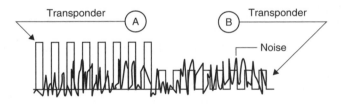

Figure 9.11   Presence of collisions in deterministic management

Here again, the question is one of knowing whether a possible collision is present, and, as before, one of the standard methods of answering this question is to measure the noise level, quantify it, weight it, and so on, in the absence of the transponder response, in order to know how to manage it during collisions.

Probabilistic collision management mode   Another approach is to use the probabilistic collision management method based on the principle of "time slots". *Figure 9.12* shows what may happen in this case.

Where this collision management method is adopted, some designers avoid the background problem by proposing the use of transponders with internal magnetic field strength measuring devices, using short periods during the time slots to transmit binary information (in a few bits, generally 2 or 3) to the base station about the strength of the field surrounding them. The base station can thus draw up an outline map of the

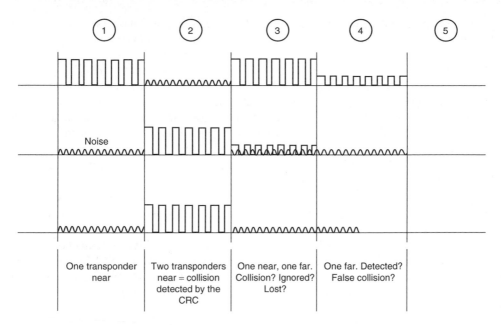

Figure 9.12 Presence of collisions in probabilistic management

transponders (whether near or far), and initially process all the transponders subjected to a strong field; it then "puts them to sleep", and, in a second step, interrogates all those subjected to a weak field – although the magnetic field has to be strengthened at this point without infringing the regulations, which is not a simple matter.

Also, if this is the case, the transponder(s) located nearest to the base station antenna will have to be put into stand-by mode, to avoid exceeding the maximum power that they can withstand without starting to smoke! (This is quite serious; it can happen, not because the integrated circuit itself "smoulders" – it can generally withstand a high junction temperature – but, if the transponder is fitted in a paper label, this is not heat resistant, and may give off real smoke, not to mention setting fire to the cardboard box and everything else.)

In this case, there is a structural separation between the period in which the base station interrogates the far transponders and that in which it deals with those located nearby (or vice versa), which may be a nuisance in some applications, for example, those in which the transponders are constantly entering and leaving the magnetic field (as in the case of articles moving on a fast conveyor).

9.5 The Demodulators

As a general rule, any good electronics textbook will give you all the necessary information about choosing the principles to be used (amplitude, phase, frequency or amplitude and phase demodulators, synchronous demodulators, digital demodulators, etc.), their qualities and defects, and the associated components.

Having said this, in contactless devices, because of the large number of parameters that have to be changed for any application (distance between the base station and transponders, coupling coefficient, etc.), it is not feasible, as I have pointed out in an earlier book, to demodulate the return signal from the transponder with an envelope or peak demodulator because this method leads, sooner or later, to the appearance of "zero lines", and thus to severe restrictions on the "safe operating areas" (SOA). To overcome these problems ("gaps" in coverage), I also mentioned that it was preferable to use synchronous or possibly quadrature demodulators.

These demodulators are so complex (in terms of the number of transistors used as well as the technology) that they are only available on the market in the form of integrated circuits, as solutions in the form of individual components are often poorly suited to applications and/or not easily produced (because of the large surface area, high cost, load performance, etc.).

One example, for 125 kHz applications, is the HT RC 110 integrated circuit that includes an "AST" analogue sampling synchronous demodulator, with two channels, I and Q, in quadrature.

For systems operating at 13.56 MHz, various recently produced circuits conform to ISO 14 443 (MF RC 500, 530, 531, for example) and ISO 15 693 (SLI RC 400) – see the table below; these also have more advanced digital synchronous demodulators in quadrature, using bit correlators.

	MF RC 500	MF RC 530	MF RC 531	SLI RC 400	CL RC 632
ISO 14 443	▓	▓	▓		▓
Part A	▓	▓	▓		▓
Part A *High speed*		▓	▓		▓
Parts A & B *High speed*			▓		▓
ISO 15 693				▓	▓

For correct detection of the bits sent from the transponder, this type of demodulator extracts the subcarrier signal present in the incident signal. To do this, the core of the quadrature demodulator uses two clocks, I_{clock} and Q_{clock}, phase shifted by $90°$ with respect to each other. The signals due to the presence of the subcarrier are then amplified, filtered and sent to a correlation circuit. The results of the correlation operation are evaluated, digitized and transmitted to the digital part of the circuit for further processing.

In the following sections, I shall just detail some further technical points relating to the thresholds and the maximum voltages to be applied to the demodulator.

Selectivity of the antenna circuit   Theoretically, the antenna circuit formed by the series or parallel LCR tuned circuit has a quality factor Q, which is necessarily associated with two important parameters, namely, the bandwidth and the pulsed response of the antenna circuit, described in detail previously. This also means that the tuned antenna circuit clearly forms a band-pass filter for both transmission and reception.

Transmission   In ASK transmission, owing to the frequency transposition, the spectrum of the baseband modulating signal (with a voltage close to the square signal) is markedly changed in terms of the current flowing in the antenna, and very different, depending on the bandwidth of the tuned circuit, from what it would have been if the antenna had had a very wide bandwidth.

Let us look at the example of 100% ASK modulation carried out by means of a pure square voltage (signal).

In the baseband, the Fourier analysis of this signal shows that it consists of a series of sinusoidal signals whose frequencies are odd-numbered multiples of order $n = 1, 3, 5, 7, \ldots$ of the fundamental frequency of the original signal, and whose fundamental frequency amplitude is greater by a factor $4/\pi$ than the amplitude of the initial square signal (all the other harmonics contribute to the reduction of this peak value, finally producing 100% of the maximum amplitude of the initial square signal – see Chapter 3).

It will therefore be necessary to take the correction factor $4/\pi$ into account when estimating the maximum current to be supplied by the output stage of the base station.

When applied to the tuned circuit of the antenna, because of the high Q ($Q > 10$) and the large associated bandwidth ($Bp @ -3\,\mathrm{dB} = f_0/Q$), most of the upper harmonics of orders (3), 5 and 7 will rapidly fall outside the band and will therefore be highly attenuated. The energy of the square signal is then principally (almost exclusively) concentrated in the first harmonic or harmonics of the spectrum.

Reception   When the signal from the transponder is received, the selectivity of the base station antenna circuit must be such that it can pass the whole range of its bandwidth.

Generally, the return signal is formed by a load modulation process including a subcarrier frequency, with the main aim of helping with collision management and improving the signal to noise ratio. It is therefore highly advisable to recover the energy contained in the sidebands of the return signal spectrum, in order to obtain the maximum benefit from the energy contained in these bands.

ISO 14 443 and ISO 15 693 indicate that these bands are located, respectively, at $fc/16 = 847.5\,\mathrm{kHz}$ and $fc/32 = 423.75\,\mathrm{kHz}$ on either side of the carrier frequency $fc = 13.56\,\mathrm{MHz}$.

Additionally, where transponders in stacks (in batches) have to be dealt with, their tuning is generally shifted towards frequencies above the carrier (in the region of $17\,\mathrm{MHz}$), which contributes to an intrinsic asymmetry between the lower sideband and the higher sideband of the subcarrier.

We must therefore determine a specific reception selectivity for each new project, in order to meet the requirements of each type of application – or, alternatively, we must produce more highly developed demodulation systems, using DSP incorporating DFT (Discrete Fourier Transform) calculations, not to be confused with FFT (Fast Fourier Transform), to estimate the energies contained in the sidebands. To resume, *Figures 9.13* and *9.14* give respectively the global selectivity of the transmission chain and some examples of signals received.

9.6   The Common Reception/Transmission Part

This part, which is common to many functions of the base station, is generally built around a standard microcontroller.

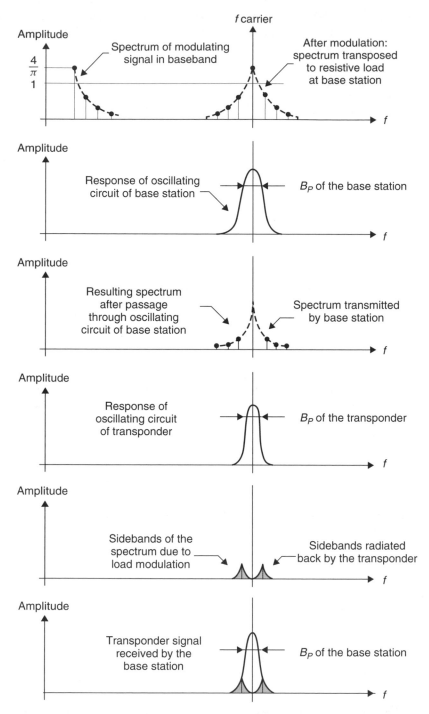

Figure 9.13 Global selectivity of the transmission-reception chain

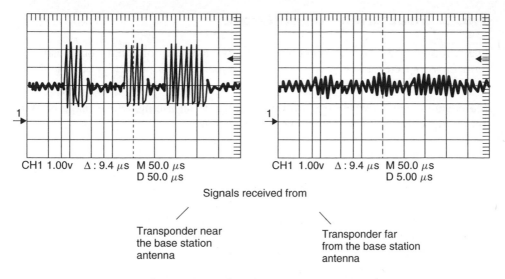

CH1 1.00v   Δ : 9.4 μs   M 50.0 μs   D 50.0 μs

CH1 1.00v   Δ : 9.4 μs   M 50.0 μs   D 5.00 μs

Signals received from

Transponder near
the base station
antenna

Transponder far
from the base station
antenna

Figure 9.14   Example of signals received

9.6.1 The microcontroller

The microcontroller in the base station generally has three main functions:

— managing the communications protocol, including, if necessary,

— collision management,

— authentication between the base station and transponder(s),

— encryption of the communications between the base station and the transponder or transponders;

— matching the planned application;

— establishing communications (RS 232, for example) with a host system, if necessary.

Most applications are supported by any microcontroller (4-bit, 8-bit, CISC, RISC).

As regards the performance of these components, the most critical points relate to their calculation speed, because with some communications protocols (in the downlink, from the transponder to the base station) the designer has to carry out some complicated manoeuvres and use large time shifts in order to correctly detect the bits arriving from the transponder, or to provide appropriate return times.

The same applies when the microcontroller has to run an encryption algorithm in a specified time to avoid a time out, which would be catastrophic for the application. Note that most of these algorithms require Boolean calculations, and some microcontrollers do not have instructions enabling them to execute these operations rapidly at bit level.

9.6.2 Structure of the software

The writing of the software programmed into the microcontroller must obviously conform to the specifications of the required communications protocols, but we should also look at its content and its structure.

The content   This is where the software developer can use his whole range of skills to provide software for the following main functions:

— management of the communications protocols, which may be simple or more complex, and must conform to the contactless standards, such as ISO 14 443 – 2/3/4, ISO 15 693 – 2/3, ISO 11 784/85 and ISO 14 223. Clearly, in the course of development, each designer will provide his own software implementation, ensuring that he has correctly interpreted the official texts, and has correctly read between the lines of these texts in order to provide all the potential compatibility and interoperability offered by these standards;

— collision management, including in all cases the "weak collisions" as described in the preceding sections;

— development of encryption algorithms with respect to performance, calculation time and byte size;

— and so on;

— and finally, the management of all the detailed requirements of the application.

The structure   As mentioned several times, the members of the standardization and design and development teams for this kind of system and components have long been keen defenders of the well-known seven-layer ISO/OSI communications model.

In applications, the seven layers of the OSI model are frequently "squashed" into three or four layers, usually the physical layer (1), the data link layer (2) and the application layer (7).

To ensure upward and downward compatibility (for contact and contactless smart cards, for example), we strongly advise designers to structure their software according to the OSI layer division, for example, on the basis of the state diagrams of the transponders used (see the author's previous book on RFID, if necessary), in order to avoid the all too frequently encountered "spaghetti" software structure.

I hope this information will make it easier for you to design your systems. I will now recommend some useful development tools.

Part Six

Tools and Measurement Methods

RFID and Contactless Smart Card Applications   D. Paret
© 2005 John Wiley & Sons, Ltd

10

Development Aids and Tools

In this penultimate chapter, I shall outline the aids and tools required for the satisfactory development of an application. By "satisfactory", I mean "according to the specifications, reproducible, reliable, secure, and according to the radiation, electromagnetic pollution and human safety standards, and so on".

To achieve this, the use of certain procedures or equipment is, if not necessary, then at least strongly recommended if you wish to avoid disasters (return of equipment, retrofitting, upgrading, etc.), which will be both costly and difficult to deal with.

Clearly, all this entails a degree of expense, which will have to be evaluated in relation to the investment, the infrastructure to be installed, the number of systems to be produced, and so on, but it must be considered in all cases.

10.1   Simulation of the System Performance

Before moving on to the hardware development, it is often useful to simulate the base station/transponder/environment system for the application, if possible.

10.1.1   Simulation models

Although the idea of using simulation models comes to mind first, it is often (extremely) difficult to construct simulation models for these systems, because they are highly complex, especially those relating to the immediate metallic and magnetic environment. Furthermore, most integrated circuit manufacturers are extremely secretive about the electronic architecture of the analogue input and communication interfaces of their transponders.

In short, whenever possible you should try to set up physical, electrically and mathematically equivalent models to make the associated software run on conventional platforms, such as PCs, and thus obtain an initial idea of the required performance of the planned system.

RFID and Contactless Smart Card Applications   D. Paret
© 2005 John Wiley & Sons, Ltd

The word "models" makes us think of validity, limits, restrictions, and so on – so proceed with caution, because while a model may be simple to describe, this very simplicity may mean that it cannot be used in a real situation (or may just be totally useless – which is why it is a good idea to publish it! Seriously – this often happens!). So you should always be careful, because these models can often provide a simulation based on an ideal world, which is sadly very unlike the daily reality of applications.

10.1.2 Simulation tools

As part of their technical support service, transponder manufacturers usually offer their major customers the software for equivalent mathematical models of their products, which can easily be integrated into higher-level software (such as MathLab or PSpice, and many others) or special-purpose tools and CAD systems.

These tools can be used for much of the initial spadework for the planned application. In this phase, it is usually possible to complete 70 to 80% of the project – but, as everybody knows, it is the remaining 20 to 30% that takes the most time!

Examples:

— For physical and mathematical reasons, it is very difficult to precisely quantify and model what happens in the presence of very weak or very strong coupling, which is most annoying in the case of dedicated systems using contactless smart cards in stacks in a wallet, or long-range applications where some transponders may be very near to the base station antenna while others are very far away.

— How can we model ALL the positions (in three dimensions) of ALL the transponders that may be positioned in a volume of several cubic metres for batch reading, and be 100% certain (not 98–99% !) that we can communicate with each one?

10.1.3 Correlation

Also, when the simulations have been completed, it is necessary to establish precise correlations between these simulations and the actual measurements made, if only in order to enrich the simulation models (which we would hope to be able to use in a future application that may have nothing in common with the previous one, and for which the results will be of very little use! But who knows?).

The measurements to be used for these correlations can only be made with special development aid tools.

10.2 Development Aid Tools

To obtain a correct picture of an application, it is necessary to define precisely the parameters of the antennae of the base station and transponder, and to know the value of the coupling coefficient and the mutual inductance between the base station and transponder.

10.2.1 Some theoretical considerations

Before examining the measurement method and the development aid tools required, let us quickly run through the definition of these two parameters.

Mutual inductance and coupling coefficient
Mutual inductance The parameter M represents the "mutual" effect (the mutual effect of a primary voltage U_1 on a secondary voltage U_2 and vice versa). The operation of this mutual inductance can be likened to the equivalent of the physical presence of an inductance M connected between the primary and secondary electrical circuits. Its unit of measurement is the henry.

Perfect mutual inductance To begin with, let us examine the case in which all the flux produced by the primary coil passes through the secondary coil (*Figure 10.1*), as in the case of a conventional ideal transformer.

If all the flux produced by the primary coil passes through the secondary coil, we say that the mutual inductance M is perfect. In this case, it is an easy matter to quantify the magnetic mutual inductance M by means of non-tuned coupled circuits (see *Figure 10.2*):

— the flow of I_1 induces an emf $= -jM \cdot \omega I_1$ in the secondary;

— the flow of I_2 induces an emf $= -jM \cdot \omega I_2$ in the primary.

According to the circuit laws applied to the primary and secondary circuits, we can state that

— in the primary, $E - jM \cdot \omega I_2 = I_1(R_1 + jX_1)$ where $X_1 = L_1 \cdot \omega$;

— in the secondary, $-jM \cdot \omega I_1 = I_2(R_2 + jX_2)$ and $X_2 = L_2 \cdot \omega$.

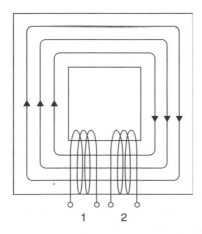

Figure 10.1 Case of a standard ideal transformer: all the flux generated by the primary coil passes through the secondary

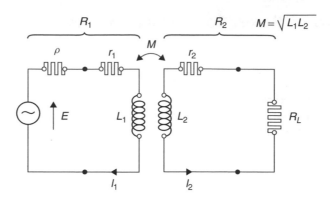

Figure 10.2   Non-tuned coupled circuits

If we subtract I_2 from the equation for the secondary and carry its value into the equation for the primary, after expansion we find:

$$E = \left[I_1 \left(R_1 + R_2 \frac{M^2 \cdot \omega^2}{R_2{}^2 + X_2{}^2} \right) \right] + \mathrm{j} \left[X_1 - X_2 \frac{M^2 \cdot \omega^2}{R_2{}^2 + X_2{}^2} \right]$$

For this complex equation to be valid, its imaginary part must be zero, that is,

$$X_1 = X_2 \frac{M^2 \cdot \omega^2}{R_2{}^2 + X_2{}^2}$$

If R_2 is small with respect to $L_2 \cdot \omega$ (which is always the case), we can ignore $R_2{}^2$ with respect to $X_2{}^2$. Thus we find that

$$L_1 \cdot \omega = \frac{M^2 \cdot \omega^2}{L_2 \cdot \omega}$$

that is, finally,

$$M = \sqrt{L_1 L_2}$$

and the maximum energy transfer takes place when

$$\frac{R_1}{R_2} = \frac{L_1}{L_2}$$

The coupling coefficient k   Now let us examine the most common case, in which only some of the flux produced by the primary coil passes through the secondary. *Figure 10.3* shows how the total flux (Φ_{total}) produced by the coil L_1, through which the current I_1 passes, is divided into two, between a useful flux (Φ_{useful}) passing through L_2 and a leakage flux (Φ_{leak}).

This phenomenon can be quantified by introducing a "coupling" parameter between L_1 and L_2.

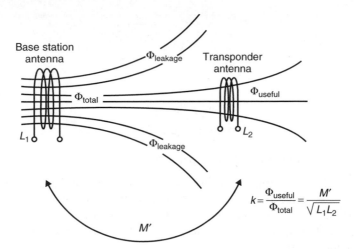

Figure 10.3 The most frequent case, in which only some of the flux generated by the primary passes through the secondary

This coupling, representing the whole assembly of components, can be "close" or "loose", and its strength clearly depends on the shapes, materials, distances, magnetic environments, and so on.

Assuming that

N_1 is the number of turns of the base station antenna,
N_2 is the number of turns of the transponder antenna and
s_2 is the surface area of one turn of the transponder antenna,

then, in terms of magnetic flux, we can say that, for the new value of mutual inductance M':

$$M' \cdot I_1 = N_2 \cdot \Phi_{21} = N_2(B_d \cdot s_2)$$

enabling us to calculate $M' = f(r, d)$:

$$M' = N_2 \left[\left(\mu \frac{r^2}{2(r^2 + d^2)^{3/2}} N_1 \right) s_2 \right] \Rightarrow M' = f(r, d)$$

In the latter case, the mutual inductance M' is no longer perfect, and its value is smaller than the value of M found in the previous section, in fact it is $x\%$ of its perfect maximum. The correction factor is called the coupling coefficient k (reflecting the extent of the coupling), which, by definition, is as follows:

$$k = \frac{M}{M'} = \frac{M'}{\sqrt{L_1 L_2}} = g(r, d) \text{ where } k \text{ is a percentage,}$$

or alternatively

$$k = \left(\mu \frac{r^2}{2(r^2 + d^2)^{3/2}} \right) N_1 N_2 s_2 \frac{1}{\sqrt{L_1 L_2}} = g(r, d)$$

$$k = \left(\mu \frac{r^2}{2(r^2 + d^2)^{3/2}} \right) s_2 \sqrt{\frac{N_1^2 \cdot N_2^2}{L_1 L_2}} = g(r, d)$$

where

$$L_1 = L_{0^1} \times N_1^2$$

$$L_2 = L_{0^2} \times N_2^2$$

where L_{0^1} and L_{0^2} are the inductances per turn.

Estimating and measuring the value of k   Using the above equations, it is not difficult to calculate the precise value of M' and therefore of k once we know the physical data (distance, etc.), but this proves to be much more difficult if we decide to introduce into the equation the magnetic parameters (metal parts, magnetic screens, etc.) of a medium surrounding the system.

Therefore, in order to avoid these difficulties, I will now describe a simple method of obtaining a very good estimate of these values by means of a measurement that is very easy to make.

For this purpose, let us return to *Figure 10.4*, which shows the non-tuned coupled circuits.

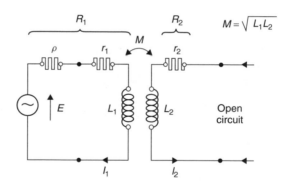

Figure 10.4   Non-tuned coupled circuits

When the secondary is not under load (no current I_2 and therefore no physical effect of the flow of the current I_2 in the primary, therefore $V_2 = V_{20}$), we can say that

$$z_1 = R_1 + jL_1 \cdot \omega$$

$$i_1 = \frac{V_1}{z_1}$$

$$v_{20} = -jM'\omega I_1$$

Taking the moduli of the complex values, we find

$$V_{20} = M' \cdot \omega I_1$$

$$= [k\sqrt{L_1 L_2}]\omega I_1$$

and, given that $I_1 = V_1/Z_1$, that is,

$$I_1 = \frac{V_1}{\sqrt{R_1{}^2 + (L_1 \cdot \omega)^2}}$$

Assuming that $Q_1 = (L_1 \cdot \omega)/R_1$ and is greater than 10 (which is always true of transponder applications), in other words, $R_1{}^2 << (L_1 \cdot \omega)^2$, we obtain

$$I_1 = \frac{V_1}{L_1 \cdot \omega}$$

and therefore

$$V_{20} = [k\sqrt{L_1 L_2}]\omega \left(\frac{V_1}{L_1 \cdot \omega} \right)$$

therefore

$$k = \frac{|V_{20}|}{|V_1|} \cdot \frac{\sqrt{L_1}}{\sqrt{L_2}}$$

where V_1, V_{20}, L_1 and L_2 are parameters that can be very easily measured in applications.

Figure 10.5 shows the type of circuit that can easily be set up to make this measurement.

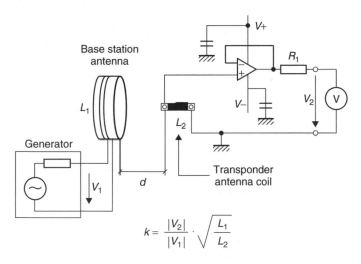

$$k = \frac{|V_2|}{|V_1|} \cdot \sqrt{\frac{L_1}{L_2}}$$

Figure 10.5   Measuring k

10.2.2   Tool for measuring the coupling coefficient

As we have shown above, we only need two physical antennae, namely, that of the base station and that of the transponder, to estimate the coupling coefficient between the antennae of the system; so, theoretically, we need no special tools, and the preceding sections are unnecessary ... unless ...

In fact,

— where the base station antenna is concerned, it is relatively easy to achieve this, since it is designed by you, the system designer;

— but for the transponder antenna the situation is quite different. Everything depends on the physical form in which it is supplied, because, if the chosen transponder is a fully overmoulded finished product, it is often very difficult to access the terminals of its antenna without damaging them by making the measurements.

In this case, to help system designers, transponder manufacturers can often supply – as a "development tool" – the same physical transponders containing only the antenna coils with two outlets ("moulded coils"), to facilitate measurements relating to the coupling coefficient of the systems, thus enabling the mutual inductance to be determined.

NOTE

In contactless applications, the orders of magnitude of the coupling coefficient k vary from a few tenths of 1% (0.3 to 0.5%) to several tens of 1% (15 to 20%). For most systems, the value is of the order of a few %.

10.2.3   Bond out chip for estimating the energy transfer

To discover whether the transponder will really be remotely supplied by the magnetic field provided by the base station antenna in the real application, it is sometimes necessary to measure not only the induced voltage present at the terminals of the integrated circuit V_{ic}, but also the current that it consumes. For this purpose, we must have access to the transponder chip encapsulated in a casing having a number of terminals (a DIL-8 housing, for example) called a "bond out chip", to facilitate the measurement of its current consumption or the supply voltage *vdd* actually present at its terminals.

10.2.4   Demonstration kits

Most manufacturers of integrated circuits and/or transponders offer more or less sophisticated demonstration kits for assistance with development and/or familiarization. These generally comprise a base station (short or long range as appropriate), its antenna, different types of transponder and one or more CD-ROMs providing numerous low-level software routines ("low-level library") for quickly testing the communication protocols and for designing the base station and transponder antennae, if necessary. Example: see *Figure 10.6* (TED kit).

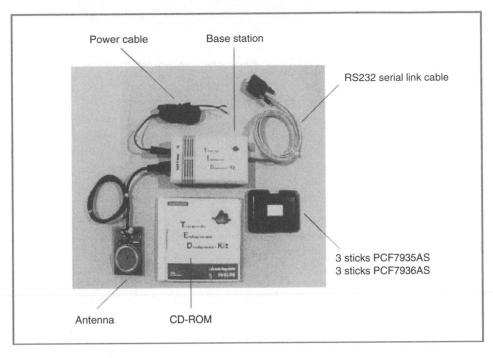

Figure 10.6   Example of a kit for demonstration, familiarization and development assistance (from a Philips SC document)

11

Measurement Methods for Contactless Systems

After this rapid survey of the available development tools, it's time to make the measurements.

We should break these down into two parts:

— measurements relating to the individual performances of the base station and the transponder, with the aim of discovering whether they conform to the desired specifications;

— measurements relating to the base station–transponder system as a whole.

11.1 The Principal Parameters to be Measured for a Transponder

To determine the characteristics of a transponder, we must define certain parameters, which we can subsequently measure, for example, the magnetic induction, the resonant frequency of the transponder, and so on.

We need to measure the parameters listed below.

11.1.1 Absorption threshold, B_{thr}

When a transponder enters the magnetic field, it changes (reduces) the field strength locally because of its presence and the energy that it absorbs for operation. Thus, there is an induction threshold B_{thr} above which the transponder starts to operate, causing energy absorption and reducing the induction (for further details, see the threshold field definition in Chapter 2).

11.1.2 Read absorption, B_{read}

Similarly, when the transponder is in read mode, it modulates its load, and the effect of this modulation can therefore be measured between the two states (loaded and unloaded), giving us the modulation depth, $B_{read}(B_{read} = B_{max} - B_{min})$.

RFID and Contactless Smart Card Applications   D. Paret
© 2005 John Wiley & Sons, Ltd

11.1.3 Resonant frequency, f_{res}

The resonant frequency f_{res} can be measured by determining the frequency at which the minimum absorption threshold B_{thr} is found (once again, you should refer to the threshold field definition in Chapter 2 for further details).

A note in passing: As mentioned several times, because of numerous application-related factors, the resonant frequency of the transponder does not have to be exactly the same as the carrier frequency delivered by the base station of a system.

11.1.4 Bandwidth

By making the measurement described above and finding the values of the curve B_{thr} as a function of the incident frequency, we can easily determine the transponder's bandwidth at 3 dB, and then deduce the exact physical value of the quality factor Q of the transponder by using the formula $Bp = f_{res}/Q$.

In the same way, we can measure the programming threshold B_{pgr} and the programming modulation index m.

The values of these parameters will clearly enable you to build up a comparative picture of the components, integrated circuits and finished transponders available on the market.

11.2 Transponder Measurement Methods and Set-ups

Although they have a strong family resemblance, we must separate the measurement methods into two classes according to the two main operating frequencies of 125 kHz and 13.56 MHz. See also ISO 18047-3 and ISO 10373-6.

11.2.1 The 125 kHz products

We need a test bench to measure these parameters.

Figure 11.1 shows the device and the arrangement of the equipment used to create a uniform magnetic field with the aid of Helmholtz coils.

This system operates as follows:

— A pair of field-generating coils, connected in series and orientated in phase magnetically, creates a uniform magnetic field between the coils. The strength of this field depends solely on the physical configuration of the coils (radius, number of turns, etc.) and the potential difference applied to them.

As shown in the figure, a pair of coils is then placed on either side of each field-generating coil. These enable the effect of the presence of the transponder to be measured as a differential amount.

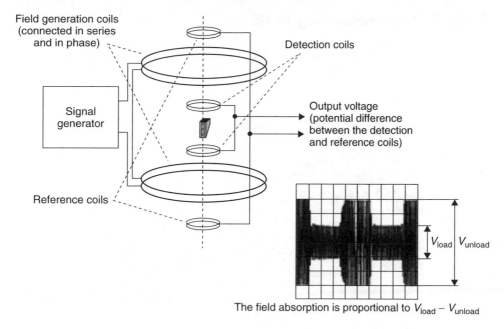

The field absorption is proportional to $V_{load} - V_{unload}$

Figure 11.1 Measurement device (Helmholtz coils) for characterizing transponders

— The first pair is the pair of reference coils. They are placed in series and in phase, and are used, as their name implies, as a reference for the measurement during the transponder modulation. Being located fairly distant from the centre of the system, they will be only slightly affected by the presence of the transponder, and the potential difference collected will act as the reference for the measurement.

— The other pair of coils, used for actually making the measurement, is the pair of sensing coils. They are also connected in series and are in phase magnetically; they are mounted in the Helmholtz configuration, in order to minimize the various changes of position of the transponders during measurement. The voltage induced across the terminals of these coils is proportional to the induction.

Since the modulation has a very weak effect on the field produced by the field-generating coils, the set of sensing and reference coils is connected in a "bridge" to provide the output signal used for measurement.

If there is no transponder in the field, the reference coils are used to minimize the output voltage, effectively by zero balancing the bridge.

When all this has been done, for a specified potential difference (of U volts) applied to the generating coils (whose impedance and quality factor Q are known), we can calculate the maximum induction B_{max} that will be available in the physical centre of the system. In the case under discussion, the relation is $B = 3.7 \times 10^{-6} \times U$ and, given the total surface area and the coupling due to the sensing coil, the sensitivity is $8.9\,\mu$ T/V.

11.2.2 The measuring tools

Numerous tools are available on the market for making these measurements. The best-known ones are made by Scemtec, see *Figure 11.2* (their address is listed at the end of this book).

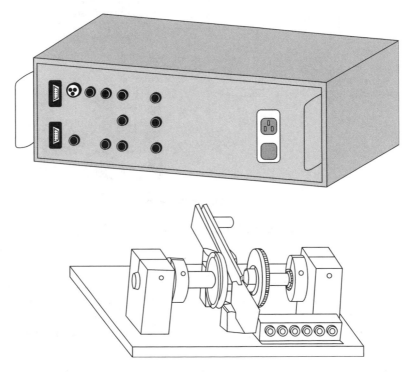

Figure 11.2 Example of a measurement bench (from a Scemtec document)

11.2.3 The 13.56 kHz products

As regards the transponders (in smart card format) and base stations (readers) according to ISO 10 536 – 2, 14 443 – 2 and 15 693 – 2, the standards of the ISO 10 373 family (−1, −4, −6, −7) describe the test tools and set-ups (calibrating coils, reference readers, reference cards), the test methods (electrostatic discharge (ESD), load modulation) and the base station test methods (communication volume, amplitude modulation, load modulation reception).

The measurement methods are very similar to those described in relation to the 125 kHz system.

By way of example, *Figures 11.3a* and *11.3b* show set-ups for testing and measuring, respectively, the power transfer and load modulation capacity between base station and transponder.

For further information on this subject, you should consult the official texts of the standards available from ISO (see address at end of book).

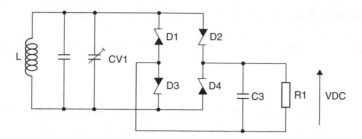

Figure 11.3a Example of a transferred power measurement device

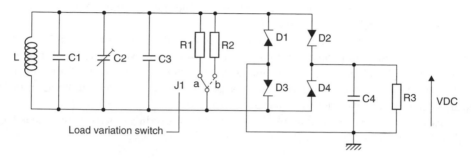

Figure 11.3b Example of a back modulation voltage measurement device

11.2.4 The measuring tools

At the present time, I am aware of only a few commercial tools (produced by Arsenal Research and Micropross – see following section) for making these measurements, and each system manufacturer usually develops his own measuring equipment according to the standards documents.

11.3 Measuring Complete Systems

Regardless of the operating frequency – 125 kHz, 13.56 MHz or other – if we wish to ensure that the system operates properly, we must carry out further sets of measurements on the systems under development.

11.3.1 The various measurements to be made

These mainly relate to the measurements of available induction according to the distances and values of the coupling coefficients k between the base station and the transponder, as mentioned in Chapter 9.

11.3.2 The measurement sequence

Should the measurements be made in a particular sequence?
   The answer is: yes!

We must start from the principle that all the inductances, the quality factors Q, the coupling coefficients, and so on, of the various components involved are known (calculated or measured), and that only the measurements of the whole system remain to be carried out.

11.3.3   Verification of the energy surfaces, the zero lines and the areas of good operation

When these measurements have been made, the designer often has to make a number of lengthy modifications to make his system reliable and reproducible, if he has not performed good preliminary simulations.

In this case, it is necessary to measure the areas of good operation, in other words, the "energy and communication areas" where you can be sure that the transponder will actually be remotely supplied, and then the equipotential lines of the received signal, according to the type and sensitivity of the base station demodulator, in order to determine whether or not there are any zero lines.

Figure 11.4 shows some examples of characteristic readings of these zero lines (see the author's previous book on RFID for more information).

11.3.4   Certification authorities

Here is yet another problem.

To understand the tasks and modus operandi of these certification authorities, we must consider some typical applications.

First case: The installation is purely proprietary; in other words, the base station and transponders are supplied by a single company which undertakes to ensure that the system operates correctly.

Second case – very common: This relates, for example, to the contact smart cards that you will be familiar with. All readers (regardless of their manufacturers) must be able to read all cards (every make of card).

The same applies to contactless technology. Clearly, to achieve this, we need to standardize. But what do the standards say about this? Sadly, the answer is not so clear.

In fact, the standards generally tell you:

—   on the one hand what a reader must be able to do in order to meet the specifications imposed by the standard in relation to a test measurement on a "reference" card;

—   on the other hand what a card must be able to do in order to meet the specifications imposed by the standard in relation to a test measurement on a "reference" reader.

At the end of such a measurement cycle, we can have, "legally speaking", two components which conform perfectly to the standard, but which may still give rise to problems in use, due to "side effects" of the specifications.

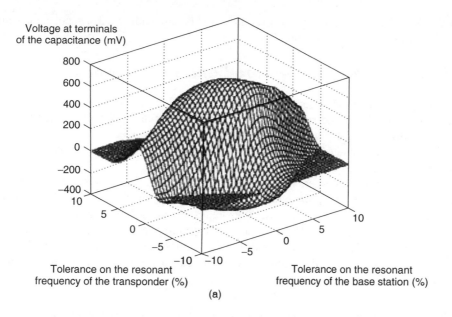

(a)

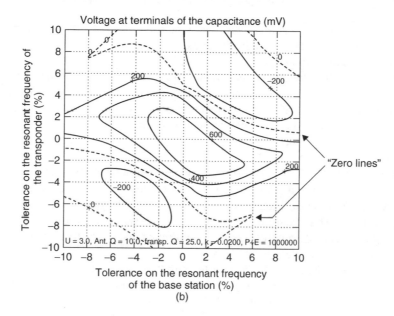

(b)

Figure 11.4 Example of "zero lines"

Therefore, for certain products for which such circumstances must be avoided in use, there are certification authorities established, not in order to test the "legal" conformity of each component with the standards, but to ensure the correct overall operation of the "reader-cards" system, regardless of the manufacturers of these components.

For example, for generic applications of the MIFARE *(trademark)* type, according to ISO 14 443 type A (for all manufacturers of integrated circuits and transponder card embedders), the Arsenal company (address in the appendix) carries out certification for all kinds of readers and cards.

11.3.5 Conformity with radiation regulations

I have already mentioned several times that contactless applications must conform to the current standards and regulations on emitted levels and limits of radiated spectra, according to the country where the systems are to be operated.

The principal standards governing these applications are currently those produced by ETSI and FFC, as cited frequently above.

Theoretically, all workers in this field comply with these standards. There is a major anomaly, however, sometimes amounting to an infringement of the Civil Law provisions relating to "Selling"; the advantages of equipment may be praised to the skies in order to achieve sales in a particular country, even though such spectacular performance is only authorized in a different country and not in the one in which the promotion is conducted; this leads only to increased problems and confusion in the market in question.

11.3.6 Human exposure to electromagnetic fields

Clearly, we must also comply with the numerous standards relating to the problems of human exposure to electromagnetic fields. Note that these standards also vary somewhat from one country to another.

One of the best-known organizations concerned with standardizing human exposure to electromagnetic fields is indisputably the International Commission on Non-ionizing Radiation Protection (ICNIRP), the successor of the International Radiation Protection Association (IRPA) and the International Non-ionizing Radiation Committee (INIRC). To give you a final abbreviation, this distinguished body regularly meets at the World Health Organization (WHO).

Essentially, ICNIRP issued some major recommendations in 2002 for signals up to 300 GHz, specifying that exposure (for 6 min) must not exceed, for the head and trunk, a Specific Absorption Rate (SAR) of 2 W/kg per 10 g of human tissue. In the United States, an ANSI standard of 1992 and an IEEE standard of 1999 stipulated 1.6 W/kg per 1 g. I provided some details of the complex nature of these measurements in my first book on contactless technology.

Generally, it is the long-range systems that are most critical in terms of compliance with these standards, since they need the highest radiated power (EIRP).

Coming down to earth after all these generalities, you should note that very few companies have the resources for correctly carrying out the measurements required by the various standards, and only a few specialist laboratories can do the job – at a cost that is usually rather high.

Generally, these measurements are made at the final stage, when the designer is practically sure that the planned system is (or should be) (in terms of theory, manufacturers' guarantees, etc.) in accordance with the standards.

Finally, you should also note that the content of these standards has been taken into account in the development of the proximity standard ISO 14 443 and the vicinity standard ISO 15 693, to ensure that systems conforming to these standards also conform to the human exposure standards.

To give you an idea of the numbers involved, here are a few orders of magnitude of the SAR, in W/kg per 1 g and per 10 g of human tissue, which you can also find in Chapter 6 on examples at 13.56 MHz:

In proximity applications (ISO 14 443): For a power in watts of approximately 0.6 W, in other words, a radiated power of approximately 20 μW:

Measuring range	SAR per 1 g	SAR per 10 g
5 mm	≈0.16	≈0.1
15 mm	≈0.08	≈0.05

In vicinity applications (ISO 15 693): For a power in watts of approximately 4 W, in other words, an authorized radiated power of approximately 10 mW:

Measuring range	SAR per 1 g	SAR per 10 g
5 mm	≈1.3	≈0.8
15 mm	≈0.7	≈0.5

This shows that, even in the most unfavourable applications (long-range systems and measurements made near the antenna, at 5 mm), these levels conform to even the strictest American standards.

11.3.7 A little more work

We haven't quite finished yet; note that the heading of this section applies to you, the reader, as much as to myself, the author, because, as you will see, there are still a few more tasks to be added to our schedule. So here we go. . .

Protection from light and X rays The transponders used in contactless devices are essentially components with very low power consumption, and therefore they use advanced technology. These technologies, including CMOS technology, small line width, E2PROM memory, low-voltage circuits, and so on, are sensitive to radiation in the form of visible and invisible light, X rays, and so on, which theoretically contribute energy and help the electrons to jump from the valence band to the conduction band of the semiconductor used. This is particularly marked in relation to the performance of analogue circuits and the storage of data in the E2PROM memory elements.

Visible light Visible light (wavelengths from approximately 400 nm (blue) to 850 nm (red and near infrared)) is therefore harmful to the integrated circuits of transponders. Manufacturers' specifications frequently contain wordings, such as "measurements have shown that light radiation of $E_{\max} = 60 \, \text{W/m}^2$ in the visible spectrum reduces the operating range of the product".

Since the energy present on a fine summer day is of the order of $260\,\text{W/m}^2$, designers and users are strongly advised to check whether or not the integrated circuit used in the transponder is protected from this radiation by means of encapsulation or coating with appropriate opacity, in other words, sufficient to provide a reduction of $260\,\text{W/m}^2$ to $30\,\text{W/m}^2 \approx 9$, that is, approximately $20\,\text{dB}$ at the wavelengths in question.

Invisible light   UV radiation causes the same problem and is remedied in the same way, during the manufacture of inlets as well – except that the same protective products cannot be used.

You may find this amusing, but one day you might be very disappointed to find that your money had disappeared from your electronic wallet (because of an erased memory) after leaving it out in the sun, on the beach or on the ski slopes! The plastic used for credit cards is surprisingly transparent to sunlight. If the designers haven't UV-protected your smart cards, don't forget to apply some maximum factor sun cream to your favourite cards to avoid sunstroke before you stretch out on your beach towel!

X rays   Fortunately, integrated circuits are generally fairly insensitive, or much less sensitive, to low-intensity X rays – but you should still be cautious!

Having reached this stage at last, you can (almost) afford to relax in the awareness that you finally know everything about your system!

11.4   Conclusions

So we come to the end of our journey through the elements of the implementation of identification devices, RFID and contactless smart card applications; I hope you have enjoyed the trip.

As you will have realized, this is a complicated field, because most of the parameters involved combine with and overlap each other, and one of my aims has been, once again, to try to straighten out this tangled web.

I hope that I have succeeded in demystifying most of the problems involved in moving from theory to practice in these applications, and that you will now find it easy to deal with the specification or implementation of a projected application. If so, I will have succeeded.

Two final remarks: I have not dealt in any depth with matters of software implementation and cryptography, as these subjects are very much orientated towards specific products and their description falls outside the scope of this book. Those of you who are interested may like to know that there are many specialist publications on the market, which lie outside the scope of this collection of titles. Readers interested in these areas should consult these publications.

Also, I have not provided any detailed discussion of the contactless identification principles and applications using UHF frequencies (approximately $870\,\text{MHz}$ in Europe and $915\,\text{MHz}$ in the United States) and Super High Frequency (SHF) or microwave frequencies ($2.45\,\text{GHz}$). Although some proprietary systems are running perfectly well, the state of standardization (band harmonization, authorized power (EIRP) harmonization, etc.) is finished, and will need more time to become fully established. But have no fear, I shall return to these matters in good time. Meanwhile . . ."wait and see!"

11.5   The Future

Before we part company, I have described many possible applications in this book, and the people I meet often ask me, "what about the future?"

Briefly, in addition to contactless smart cards, labels, immobilizers, and so on, as mentioned above, workers in this field are looking towards applications including intelligent "microsystems" with "on-board electronics" and microcontrollers, capable of controlling remote sensors of all types, remotely supplied and incorporated in other systems (mechanical stress measurements for concrete pillars, intelligent air bags, tyre pressure and temperature measurement, etc.). These devices generally demand a greater or lesser amount of computing power, and therefore have higher consumption levels and require technological advances to resolve the problems of remotely supplying systems via contactless links.

The same applies to the development of "multichip" devices incorporating the protected communication section and remote supply and other more complex functions.

Clearly, the future awaits us!

Meanwhile, I hope we can meet again for further explorations of other emerging fields. So, good-bye for now.

Appendix A

A.1 Duality of Series and Parallel RLC Circuits

See the following pages.

RFID and Contactless Smart Card Applications   D. Paret
© 2005 John Wiley & Sons, Ltd

A.1.1 Series RLC resonant circuit

Power supply The series RLC circuit is supplied by a sinusoidal voltage source $E = E$m.

To avoid modifying the total resistance of the circuit, the source is considered to have zero internal resistance. This is a constant voltage supply (see *Figure A1.a*).

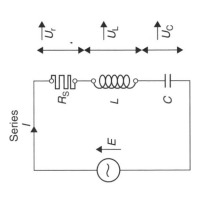

Series

Figure A1.a

A.1.2 Parallel RLC resonant circuit

Power supply The parallel RLC circuit is supplied by a sinusoidal current source $I = I$m.

To avoid modifying the total resistance of the circuit, the source is considered to have infinite internal resistance. This is a constant current supply (see *Figure A1.b*).

Parallel

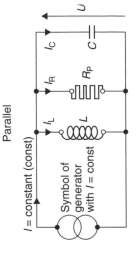

Figure A1.b

Impedance and phase shift Since the voltage is the same at the terminals of the three elements, let us consider V to be the origin of the phases and plot the current diagrams (see *Figure A1.d*).

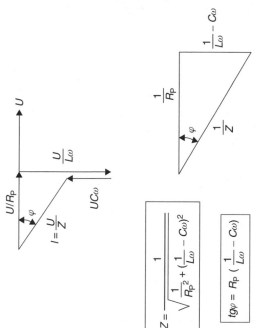

$$Z = \frac{1}{\sqrt{\dfrac{1}{R_P^2} + \left(\dfrac{1}{L\omega} - C\omega\right)^2}}$$

$$tg\varphi = R_P \left(\frac{1}{L\omega} - C\omega\right)$$

Figure A1.d

Dividing the three sides of the triangle thus formed by V, we obtain the apparent impedance triangle.

Impedance and phase shift Since the current is the same in the three elements, let us consider I to be the origin of the phases and plot the voltage drop diagrams (see *Figure A1.c*).

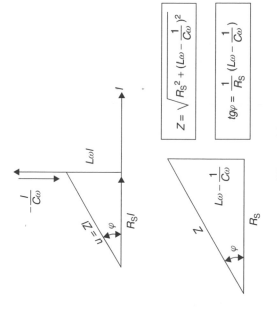

$$Z = \sqrt{R_S^2 + \left(L\omega - \frac{1}{C\omega}\right)^2}$$

$$tg\varphi = \frac{1}{R_S}\left(L\omega - \frac{1}{C\omega}\right)$$

Figure A1.c

Dividing the three sides of the triangle thus formed by I, we obtain the apparent impedance triangle.

Variations of impedance as a function of frequency The variations of impedance as a function of frequency are shown in *Figure A2.b*:

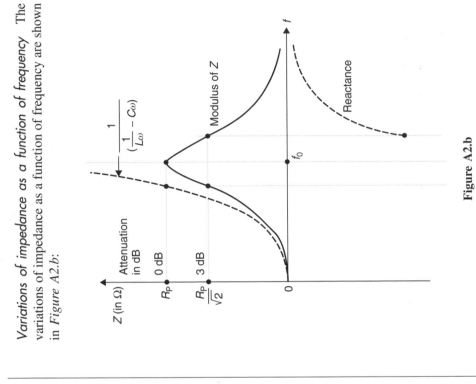

Figure A2.b

Variations of impedance as a function of frequency The variations of impedance as a function of frequency are shown in *Figure A2.a*:

Figure A2.a

— at zero frequency, $Z = 0$;

— at the particular frequency $f_0 = 1/(2\pi\sqrt{LC})$, the curve passes through a maximum $Z = Rp$;

— at infinite frequency, $Z = 0$.

Phase shifts as a function of frequency

The phase shifts are shown in *Figure A2.d.*

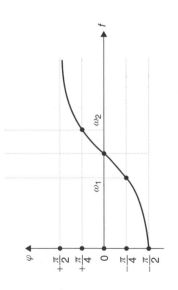

Figure A2.d

— at zero frequency, $Z = \infty$;

— at low frequency, this curve is asymptotic from $1/(C\omega)$;

— at the particular frequency $f_0 = 1/(2\pi\sqrt{LC})$, the curve passes through a minimum $Z = R$;

— at high frequencies, the curve is asymptotic from $L\omega$;

— at infinite frequency, $Z = \infty$;

Phase shifts as a function of frequency

The phase shifts are shown in *Figure A2.c.*

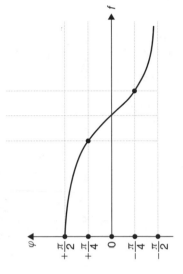

Figure A2.c

Excess voltage or quality factor

At resonance, this factor is defined as the ratio

$$Q = \frac{UL}{URs} = \frac{UC}{URs}$$

$$Q = \frac{L\omega}{Rs} = \frac{1}{Rs \cdot C\omega} = \frac{1}{Rs}\sqrt{\frac{L}{C}}$$

Selectivity and bandwidth Variation of current as a function of frequency:

$$I = \frac{E}{\sqrt{R^2 + \left(L\omega - \frac{1}{C\omega}\right)^2}}$$

At resonance, $I = (E/R)$.
The selectivity factor is, by definition,

$$S = \frac{IO}{I}$$

Excess current or quality factor

At resonance, this factor is defined as the ratio

$$Q = \frac{IL}{IRp} = \frac{IC}{IRp}$$

$$Q = \frac{Rp}{L\omega} = Rp \cdot C\omega = \frac{1}{Rp}\sqrt{\frac{L}{C}}$$

Selectivity and bandwidth

Variation of voltage as a function of frequency:

$$U = I\sqrt{R^2 + \left(L\omega - \frac{1}{C\omega}\right)^2}$$

At resonance, $UO = I \cdot Rp$.
The selectivity factor is, by definition,

$$S = \frac{UO}{U}$$

The curve in *Figure A3.a* shows the variations.

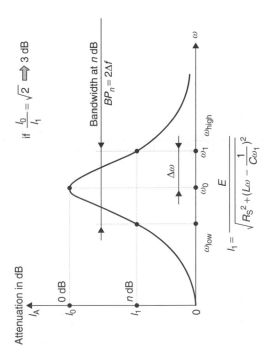

Figure A3.a   •

$$l_1 = \frac{E}{\sqrt{R_S^2 + \left(L\omega - \dfrac{1}{C\omega_1}\right)^2}}$$

Assuming that

$$\Delta f = \frac{\Delta\omega}{2\pi}$$

we can show that the selectivity is practically equal to

$$S\sqrt{1 + 4Q^2\left(\frac{\Delta\omega}{\omega_0}\right)^2}$$

The curve in *Figure A3.b* shows the variations.

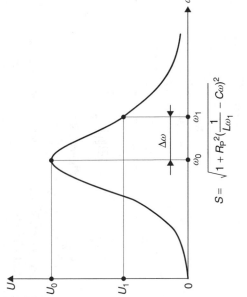

$$S = \sqrt{1 + R_P^2\left(\frac{1}{L\omega_1} - C\omega\right)^2}$$

Figure A3.b

Assuming that

$$\Delta f = \frac{\Delta\omega}{2\pi}$$

we can show that the selectivity is practically equal to

$$S\sqrt{1 + 4Q^2\left(\frac{\Delta\omega}{\omega_0}\right)^2}$$

Bandwidth at −3 dB

Assuming that $S = \sqrt{2}$, therefore 3 dB, we obtain:

$$Bp = 2\Delta f = \frac{fO}{Q}$$

Bandwidth at −3 dB

Assuming that $S = \sqrt{2}$, therefore 3 dB, we obtain:

$$Bp = 2\Delta f = \frac{fO}{Q}$$

$$\boxed{S = \sqrt{1 + 4Q^2\left(\frac{\Delta\omega}{\omega_0}\right)^2}}$$

We find the selectivity formula for a series resonant circuit, the "current selectivity", becoming "voltage selectivity" (parallel).

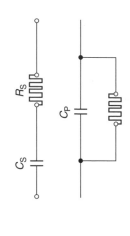

Figure A4.b

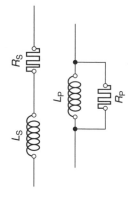

Figure A4.a

We now identify the complex impedances of the two equivalent circuits,

$$Rs - \frac{j}{Cs\omega} = j\frac{Rp}{1 + jRpCp\omega}$$

Denoting the two circuits $LsRs$ and $LpRp$,

$$Rs + jLs\omega = j\frac{RpLp\omega}{Rp + jLp\omega}$$

Introducing the quality factor of the circuit,

$$Q = \frac{Ls\omega}{Rp} = \frac{Rp}{Lp\omega}$$

$$Rs + jLs\omega = j\frac{Rp}{Q+j} = j\frac{Rp(Q-j)}{Q^2+1}$$

$$Rs + jLs\omega = \frac{Rp}{Q^2+1} + j\frac{QRp}{Q^2+1}$$

which gives us

$$\boxed{Rs = \frac{Rp}{Q^2+1}} \quad (1)$$

$$Ls\omega = \frac{QRp}{Q^2+1} \quad (2)$$

Introducing Lp, with $Lp\omega = (Rp)/Q$,

$$\boxed{Ls = Lp\frac{Q^2}{Q^2+1}}$$

$$= \frac{Rp}{1+Rp^2Cp^2\omega^2} - j\frac{Rp^2Cp\omega}{1+Rp^2Cp^2\omega^2}$$

we obtain

$$Rs = \frac{Rp}{1+Rp^2Cp^2\omega^2}, \quad Cs = Cp\frac{1+Rp^2Cp^2\omega^2}{Rp^2Cp^2\omega^2}$$

Introducing the parallel quality factor

$$Q = RpCp\omega$$

we obtain

$$\boxed{\begin{array}{c} Rs = \dfrac{Rp}{Q^2+1} \\[2ex] Ls\omega = \dfrac{QRp}{Q^2+1} \end{array}}$$

we can disregard 1 before Q^2 and,

$$Rs \approx \frac{Rp}{Q^2}, \quad Cs \approx Cp$$

A.2 Useful Addresses and Information

A.2.1 Standardization, certification and similar authorities

AFNOR

Association Française de Normalisation – AFNOR
11, avenue Francis de Pressensé
93571 Saint-Denis La Plaine CEDEX
France
telephone: + 33 1 41 62 80 00
fax: + 33 1 49 17 90 00
www.afnor.fr

ETSI

European Telecommunication Standards Institute – ETSI
Route des Lucioles
Valbonne
06921 Sofia Antipolis Cedex
France
telephone: + 33 4 92 94 42 00
fax: + 33 4 93 65 47 16

ARSENAL

Faradaygasse 3, Objekt 214
A – 1030 Vienna
Austria
telephone: + 43 1 797 47 559
fax: + 43 1 799 19 55

GENCOD EAN France

2, rue Maurice Hartmann
92137 Issy-les-Moulineaux
France
telephone: + 33 1 40 95 54 10
fax: + 33 1 40 95 54 49

A.2.2 Hardware, equipment, component manufacturers

Specialist test and measuring equipment
SCEMTEC

Scemtec Transponder Technology Gmbh
Gewerbeparkstr. 20
D – 51580 Reichshof – Wehnrath
Germany
telephone: + 49 2265 996-0

fax: + 49 2265 996-299

www.scemtec.com

MICROPROSS

33 rue Gantois

59000 Lille

France

telephone: + 33 3 20 74 66 30

fax: + 33 3 20 74 66 35

Development software   Numerous development software packages are available on the market and enable a complete system to be modelled without (too much) difficulty. Most developers use well-known software, such as SPICE or, more generally, Matlab.

Many specialized software packages for magnetic field simulation calculations are available, but unfortunately the only high-performance products relating to the RFID market (i.e. those including the real magnetic environment of the application and having a very small discrepancy – of the order of 1% – between simulation and reality) are those that have been developed with considerable effort by private companies and/or by universities commissioned by private companies; in other words, they are not available via the conventional market.

Development tools

Simulators, emulators, assemblers and compilers for integrated circuits for transponders with microcontrollers:

ASHLING

2, rue Alexis de Tocqueville

Parc de Haute Technologie

92183 Antony Cedex

France

telephone: + 33 1 46 66 27 50

fax: + 33 1 46 74 99 88

RAISONANCE

17, avenue Jean Kuntzmann

38330 Montbonnot Saint Martin

France

telephone: + 33 4 76 61 02 33

Demonstration kits: Most component manufacturers supply and/or sell demonstration or development aid kits.

Component manufacturers   Numerous component manufacturers (Philips Semiconductors, Infineon, Texas Instruments, Motorola, ST Microelectronics, Atmel/Temic, MicroChip, Inside Contactless, etc.) have been mentioned in this book.

Instead of filling a dozen pages with their addresses, fax numbers, and so on, I advise readers to visit the web sites of these large companies, which are generally highly detailed and will give you the latest news about RFID components.

Inlet and transponder manufacturers   Many companies supply semi-finished products, such as "inlets" (an inlet is an integrated circuit + antenna, mounted on a support ready for encapsulation).

A.2.3 Other useful books

In French:

Dominique Paret. *Identification radiofréquence et cartes à puce sans contact – Description* (Radio frequency identification and contactless smart cards – Description). Dunod, 2001

The basis, fundamental principles and standards of contactless technology and RFID.

In German:

Klaus Finkenzeller. *RFID – Handbuch*. Hanser, 1998

General work covering most contactless applications.

In English:

Klaus Finkenzeller. *RFID – Handbook*. John Wiley and Sons, 1999

Translation of the German-language book.

J.D. Gerdeman. *RF/ID – Radio Frequency Identification – Application 2000*. Research Triangle Consultants – RTC Inc, 1996

Oriented towards problems of logistics and road transport applications.

Index

absorption
    read, B_{read}, 303
access control, 101
AGC (Automatic Gain Control), 283
    amplifier
    balanced output, 74
    unbalanced output, 74
amplitude shift keying (ASK), 50
analogue part, 269
antenna
    3D, 261
    active, 264
    balanced, 245, 246
    coil, 223
    compensated, 245
    cubic, 261
    etched, 223
    figure of eight coil, 204, 244
    for batch reading, 260
    in aluminium tube, 252
    loaded and tuned, 29
    non-protected active, 265
    non-synchronized or multiplexed, 258
    passive, 264
    printed, 223
    protected active, 265
    remote, 74
    remote passive, 264
    screened, 245, 247
    shielded, 248
    simple circular, square or rectangular flat, 242
    synchronized, 255
    three-dimensional, 261

applications
    automobile, 261
    supermarket gate, 260
areas
    energy and communication, 308
ART, 93, 206

balun, 76, 155
balun system, 76
bandwidth at 3 dB, 304
base station, 3
batch reading, 264
battery-assisted systems, 266
Biot-Laplace law, 8
Biot-Savart law, 8
black hole, 257
BPSK, 64, 278

capacitance
    hand, 252
    parasitic, 237
    tuning, 237
capacitance bridge, 71, 155
Carson-Laplace transform, 60
certification authorities, 308
choice of operating frequency, 90
circuit
    complete equivalent transponder, 30
    natural or physical, 20
circular coil, 231
closed distance, 111
closed systems, 76
coefficient
    self-induction, 227

RFID and Contactless Smart Card Applications D. Paret
© 2005 John Wiley & Sons, Ltd